Reviews of Environmental Contamination and Toxicology

VOLUME 222

T0142859

For further volumes:
http://www.springer.com/series/398

Reviews of Environmental Contamination and Toxicology

Editor
David M. Whitacre

VOLUME 222

 Springer

Coordinating Board of Editors

ISSN 0179-5953
ISBN 978-1-4939-0248-4   ISBN 978-1-4614-4717-7 (eBook)
DOI 10.1007/978-1-4614-4717-7
Springer New York Heidelberg Dordrecht London

Foreword

International concern in scientific, industrial, and governmental communities over traces of xenobiotics in foods and in both abiotic and biotic environments has justified the present triumvirate of specialized publications in this field: comprehensive reviews, rapidly published research papers and progress reports, and archival documentations. These three international publications are integrated and scheduled to provide the coherency essential for nonduplicative and current progress in a field as dynamic and complex as environmental contamination and toxicology. This series is reserved exclusively for the diversified literature on "toxic" chemicals in our food, our feeds, our homes, recreational and working surroundings, our domestic animals, our wildlife, and ourselves. Tremendous efforts worldwide have been mobilized to evaluate the nature, presence, magnitude, fate, and toxicology of the chemicals loosed upon the Earth. Among the sequelae of this broad new emphasis is an undeniable need for an articulated set of authoritative publications, where one can find the latest important world literature produced by these emerging areas of science together with documentation of pertinent ancillary legislation.

Research directors and legislative or administrative advisers do not have the time to scan the escalating number of technical publications that may contain articles important to current responsibility. Rather, these individuals need the background provided by detailed reviews and the assurance that the latest information is made available to them, all with minimal literature searching. Similarly, the scientist assigned or attracted to a new problem is required to glean all literature pertinent to the task, to publish new developments or important new experimental details quickly, to inform others of findings that might alter their own efforts, and eventually to publish all his/her supporting data and conclusions for archival purposes.

In the fields of environmental contamination and toxicology, the sum of these concerns and responsibilities is decisively addressed by the uniform, encompassing, and timely publication format of the Springer triumvirate:

Reviews of Environmental Contamination and Toxicology [Vol. 1 through 97 (1962–1986) as Residue Reviews] for detailed review articles concerned with any aspects of chemical contaminants, including pesticides, in the total environment with toxicological considerations and consequences.

Bulletin of Environmental Contamination and Toxicology (Vol. 1 in 1966) for rapid publication of short reports of significant advances and discoveries in the fields of air, soil, water, and food contamination and pollution as well as methodology and other disciplines concerned with the introduction, presence, and effects of toxicants in the total environment.

Archives of Environmental Contamination and Toxicology (Vol. 1 in 1973) for important complete articles emphasizing and describing original experimental or theoretical research work pertaining to the scientific aspects of chemical contaminants in the environment.

Manuscripts for Reviews and the Archives are in identical formats and are peer reviewed by scientists in the field for adequacy and value; manuscripts for the *Bulletin* are also reviewed, but are published by photo-offset from camera-ready copy to provide the latest results with minimum delay. The individual editors of these three publications comprise the joint Coordinating Board of Editors with referral within the board of manuscripts submitted to one publication but deemed by major emphasis or length more suitable for one of the others.

Coordinating Board of Editors

Preface

The role of *Reviews* is to publish detailed scientific review articles on all aspects of environmental contamination and associated toxicological consequences. Such articles facilitate the often complex task of accessing and interpreting cogent scientific data within the confines of one or more closely related research fields.

In the nearly 50 years since *Reviews of Environmental Contamination and Toxicology* (formerly *Residue Reviews*) was first published, the number, scope, and complexity of environmental pollution incidents have grown unabated. During this entire period, the emphasis has been on publishing articles that address the presence and toxicity of environmental contaminants. New research is published each year on a myriad of environmental pollution issues facing people worldwide. This fact, and the routine discovery and reporting of new environmental contamination cases, creates an increasingly important function for *Reviews*.

The staggering volume of scientific literature demands remedy by which data canbe synthesized and made available to readers in an abridged form. *Reviews* addresses this need and provides detailed reviews worldwide to key scientists and science orpolicy administrators, whether employed by government, universities, or the private sector.

There is a panoply of environmental issues and concerns on which many scientists have focused their research in past years. The scope of this list is quite broad, encompassing environmental events globally that affect marine and terrestrial ecosystems; biotic and abiotic environments; impacts on plants, humans, and wildlife; and pollutants, both chemical and radioactive; as well as the ravages of environmental disease in virtually all environmental media (soil, water, air). New or enhanced safety and environmental concerns have emerged in the last decade to be added to incidents covered by the media, studied by scientists, and addressed by governmental and private institutions. Among these are events so striking that they are creating a paradigm shift. Two in particular are at the center of ever increasing media as well as scientific attention: bioterrorism and global warming. Unfortunately, these very worrisome issues are now superimposed on the already extensive list of ongoing environmental challenges.

The ultimate role of publishing scientific research is to enhance understanding of the environment in ways that allow the public to be better informed. The term "informed public" as used by Thomas Jefferson in the age of enlightenmen tconveyed the thought of soundness and good judgment. In the modern sense, being "well informed" has the narrower meaning of having access to sufficient information. Because the public still gets most of its information on science and technology from TV news and reports, the role for scientists as interpreters and brokers of scientific information to the public will grow rather than diminish. Environmentalism is the newest global political force, resulting in the emergence of multinational consortiato control pollution and the evolution of the environmental ethic. Will the new-politics of the twenty-first century involve a consortium of technologists and environmentalists, or a progressive confrontation? These matters are of genuine concernto governmental agencies and legislative bodies around the world.

For those who make the decisions about how our planet is managed, there is an ongoing need for continual surveillance and intelligent controls to avoid endangering the environment, public health, and wildlife. Ensuring safety-in-use of the manychemicals involved in our highly industrialized culture is a dynamic challenge, for the old, established materials are continually being displaced by newly developed molecules more acceptable to federal and state regulatory agencies, public healthofficials, and environmentalists.

Reviews publishes synoptic articles designed to treat the presence, fate, and, if possible, the safety of xenobiotics in any segment of the environment. These review scan be either general or specific, but properly lie in the domains of analytical chemistry and its methodology, biochemistry, human and animal medicine, legislation, pharmacology, physiology, toxicology, and regulation. Certain affairs in food technology concerned specifically with pesticide and other food-additive problems may also be appropriate.

Because manuscripts are published in the order in which they are received in final form, it may seem that some important aspects have been neglected at times. However, these apparent omissions are recognized, and pertinent manuscripts are likely in preparation or planned. The field is so very large and the interests in it are so varied that the editor and the editorial board earnestly solicit authors and suggestions of under represented topics to make this international book series yet more useful and worthwhile.

Justification for the preparation of any review for this book series is that it deals with some aspect of the many real problems arising from the presence of foreign chemicals in our surroundings. Thus, manuscripts may encompass case studies from any country. Food additives, including pesticides, or their metabolites that may persist into human food and animal feeds are within this scope. Additionally, chemical contamination in any manner of air, water, soil, or plant or animal life is within these objectives and their purview.

Manuscripts are often contributed by invitation. However, nominations for new topics or topics in areas that are rapidly advancing are welcome. Preliminary communication with the editor is recommended before volunteered review manuscripts are submitted.

Summerfield, NC, USA   David M. Whitacre

Contents

Persistence, Bioaccumulation, and Toxicity of Halogen-Free Flame Retardants

Susanne L. Waaijers, Deguo Kong, Hester S. Hendriks, Cynthia A. de Wit,
Ian T. Cousins, Remco H.S. Westerink, Pim E.G. Leonards,
Michiel H.S. Kraak, Wim Admiraal, Pim de Voogt, and John R. Parsons

Contents

Susanne L. Waaijers, Deguo Kong, and Hester S. Hendriks contributed equally to this work.

S.L. Waaijers (✉) • M.H.S. Kraak • W. Admiraal • P. de Voogt • J.R. Parsons
Institute for Biodiversity and Ecosystem Dynamics (IBED), University of Amsterdam,
P.O. Box 94248, Amsterdam 1092 GE, The Netherlands
e-mail: s.l.waaijers@gmail.com

D. Kong • C.A. de Wit • I.T. Cousins
Department of Applied Environmental Science (ITM), Stockholm University,
Stockholm SE-106 91, Sweden

H.S. Hendriks • R.H.S. Westerink
Institute for Risk Assessment Sciences (IRAS), Utrecht University, P.O. Box 80.177, Utrecht
3508 TD, The Netherlands

P.E.G. Leonards
Institute for Environmental Studies (IVM), VU University Amsterdam,
De Boelelaan 1087, Amsterdam 1081 HV, The Netherlands

D.M. Whitacre (ed.), *Reviews of Environmental Contamination and Toxicology*,
Reviews of Environmental Contamination and Toxicology 222,
DOI 10.1007/978-1-4614-4717-7_1, © Springer Science+Business Media New York 2013

1   Introduction

Polymers are synthetic organic materials that have a high carbon and hydrogen content, which renders them readily combustible. When used in buildings, electrical appliances, furniture, textiles, transportation, mining, and in many other applications, polymers have to fulfill flame retardancy regulatory requirements, primarily as mandatory specifications that often differ among countries. To achieve these requirements, chemical additives known as flame retardants (FRs) are incorporated into the polymers. In contrast to most additives, FRs can appreciably impair the material properties of polymers (United Nations Environment Programme (UNEP) 2008). The key challenge is therefore to find a suitable compromise between the performance of the polymers and fulfilling flame retardancy requirements. Brominated flame retardants (BFRs) are rather widely used because they have a low impact on the polymer's characteristics, are very effective in relatively low amounts compared to other FRs (Alaee et al. 2003), and are relatively cheap (Birnbaum and Staskal 2004). In 2004, BFRs accounted for about 21% of the total world production of FRs (SRI Consulting (SRIC) 2004). Many BFRs, however, have unintended negative effects on the environment and human health. Some are very persistent (Robrock et al. 2008), some bioaccumulate in aquatic and terrestrial food chains (Boon et al. 2002), and some show serious adverse effects such as endocrine disruption (Meerts et al. 2001). Some BFRs (polybrominated diphenyl ethers (PBDEs), hexabromocyclododecane (HBCD), and tetrabromobisphenol-A

(TBBPA), in particular) have been found in increasing concentrations in the human food chain, human tissues, and breast milk (Schantz et al. 2003; Hites 2004; Fängström et al. 2005). In 2000, exponentially increasing PBDE concentrations were measured in Swedish human milk (Norén and Meironyté 2000), and this was later followed by reports of even higher PBDE concentrations in human milk from the USA (Schecter et al. 2008).

Concerns about the persistence, bioaccumulation, and toxicity (PBT) of some BFRs have led to a ban on the production and use of many of these compounds, i.e., the hexa-, octa-, and deca-brominated biphenyls (polybrominated biphenyls or PBBs); the tetra-, penta-, hexa-, hepta-, octa-, and deca-BDEs; and HBCD (United Nations Environment Programme (UNEP) and Stockholm Convention—Press release 8 May 2009; World Health Organisation (WHO) 1994; OSPAR 2001 (2004 updated); European Parliament (E.P.) 2002; Albemarle corporation 2009; Chemtura 2009; ICL 2009). Hence, there is growing interest in substituting BFRs with alternative halogen-free flame retardants (HFFRs), and several furniture manufacturers have already voluntarily replaced BFRs with alternative HFFRs (Betts 2007).

Many HFFRs are already marketed, although their environmental behavior and toxicological properties are only known to a limited extent and their potential impact on the environment cannot yet be properly assessed. Therefore, banning BFRs and replacing them with HHFRs introduces the dilemma that little is known about the environment and human health risks of the HFFRs. Consequently, there is urgent need for information on the PBT properties of HFFRs. Therefore, the aim of this review is to make an inventory of the data that are available on the physical–chemical properties, production volumes, PBT of a selection of HFFRs that are suitable replacements for BFRs in polymers.

2 Selected HFFRs

HFFRs can be divided into several categories (see Table 1), the most important ones being inorganic flame retardants and synergists (mostly used for electronics and electrical equipment), organophosphorus compounds and their salts (housings of consumer products), nitrogen-based organic flame retardants (electronics and electrical equipment), and intumescent systems (textile coatings). From these categories, 13 HFFRs were selected for inclusion in this review as potential replacements for BFRs in polymers: aluminum trihydroxide, magnesium hydroxide, zinc borate, zinc hydroxystannate and zinc stannate (inorganic flame retardants and synergists); aluminum diethylphosphinate, bisphenol-A bis(diphenylphosphate), 9,10-dihydro-9-oxa-10-phosphaphenanthrene-10-oxide (or dihydrooxahosphaphenanthrene), resorcinol bis(diphenylphosphate) and triphenylphosphate (organophosphorus compounds and salts), melamine polyphosphate (nitrogen based organic flame retardant); ammonium polyphosphate, and pentaerythritol (intumescent systems).

Table 1 Brominated flame retardant (BFR) applications and halogen-free flame retardant (HFFR) alternatives (ENFIRO 2009)

Application	Main BFR	HFFR alternatives
Printed circuit boards Electronic components encapsulations Technical laminate	Tetrabromobisphenol-A (TBBPA)	• Dihydrooxahosphaphenanthrene (DOPO) • Zinc hydroxystannate (ZHS) • Zinc stannate (ZS) • ZHS/ZS coated with aluminum trihydroxide (ATH)
Housings of electronic products Wiring parts Housings for business machines, toys, telephones, and others consumer electronics	TBBPA Decabrominated diphenyether (DecaBDE) And other BFR	• Resorcinol bis(diphenylphosphate) (RDP) • Bisphenol-A bis(diphenylphosphate) (BDP) • Triphenylphosphate (TPP)
Electrical and electronic equipment, connectors, switches etc. Encapsulated electronic components	Brominated polystyrenes and other BFR	• Aluminum diethylphosphinate (ALPI) • Melamine polyphosphate (MPP) • Anhydrous zinc borate (ZB) • ZS
Wire and cables	DecaBDE and other BFRs	• ATH • $Mg(OH)_2$ • ZHS • ZS • ZB
Textile coatings	Hexabromocyclododecane (HBCDD)	Intumescent systems consisting of Ammonium polyphosphate (APP) + pentaerythritol (PER) + MPP

3 Characteristics of the Selected HFFR

In each of the following sections, a specific group of flame retardants and their intrinsic properties is addressed. We start with their physical–chemical properties and then we present PBT data. In the toxicity paragraphs, we report ecotoxicity data as well as effects on mammals and data on in vitro toxicity endpoints. The available data are classified based on the REACH system (*R*egistration, *E*valuation, *A*uthorisation and Restriction of *Ch*emical substances), i.e., European Union REACH legislation Regulation No. 1907/2006 Annex XIII and No. 1272/2008 Chaps. 3 and 4 (European Union 2006, 2008). This means that we assigned the data categories as being "high," "moderate," and "low."

During our literature search, we preferred data published in peer-reviewed scientific papers over those in reports and other so-called grey literature. Whenever provided in the papers we found, the most relevant details are reported. The transparency of the experimental setup was of high importance; the more study detail that was provided on test conditions and results, the more reliable we considered the data to be. Although we preferred primary sources, in some cases we referred to secondary reports (trusted independent sources such as UNEP and US EPA). Therefore, when using such data reported in this review, we strongly recommend readers also consult the original reference.

Details about endpoints chosen and the classification system used are explained in the following paragraphs.

3.1 Physical–Chemical Properties

Physical–chemical properties are highly important in assessing the environmental fate and behavior of compounds. Properties of particular interest are: molecular weight (MW), melting point and temperature of decomposition, vapor pressure, water solubility, Henry's law constant (H), the air–water partition coefficient (K_{AW}, which is closely related to H), and the octanol–water partition coefficient (K_{OW}). Specific approaches for checking the consistency between different reported values of solubility in water, vapor pressure, and Henry's law constant are available, such as the three solubility approach (Cole and Mackay 2000; Schenker et al. 2005). However, we have not attempted to differentiate methods for gathering such data in this review, since HFFR data are often scattered and fragmentary. Instead, when few reliable data points were available on a compound, estimation software or on-line calculators were used to estimate values for physical–chemical properties. When software estimators were needed for organic chemicals (or chemicals acting like organics from the provisional list), tools such as COSMOtherm® Vers. C2.1, EPI Suite 4.1 and SPARC On-Line Calculator 4.5 from the US EPA (Hilal et al. 2003, 2004; Eckert and Klamt 2010; US EPA 2011) were used. To our knowledge, no tools are available for estimating the physical–chemical properties of inorganic substances. Nor, are some property descriptors relevant for describing the partitioning of inorganic substances.

3.2   Environmental Presence and Production Volumes

The environmental occurrence of the selected HFFRs was surveyed by searching the published literature. The results of this survey revealed a lack of data on environmental presence; therefore, we thought it advisable to add information on production volumes to the review. The production volumes of the selected HFFRs can be categorized as low production, import volumes (LPV), or high production volume (HPV). The HFFRs having LPV had volumes varying between 10 and 1,000 t year^{-1}, whereas those with HPV exceeded 1,000 t year^{-1} (European Union 1993).

3.3   Persistence

The persistence of the selected compounds was evaluated by collecting data on ready biodegradability and/or dissipation times. Ready biodegradability is usually determined by performing biological degradation tests (often by microorganisms from waste water treatment plant sludge) in water, according to standard OECD guidelines (Organisation for Economic Co-operation and Development (OECD) 1992). Dissipation times may be reported for air, water, sediment/soil, and sludge. Often, half-lives were given, which is the time required for 50% of the compound to be transformed. However, it was often unclear whether these half-lives represented full mineralization, oxidation, or merely primary degradation (removal of the parent compound only). Therefore, we chose to report these values as dissipation times (DT_{50}), in which the concentration is reduced to 50% of the initial concentration after a given period. If DT_{50} values were not available, DT_{x} values are reported, where x represents the converted percentage (e.g., DT_{30} means time for the concentration to dissipate to 30% of the initial concentration). Depletion processes not involving transformations, such as sorption, evaporation, and scavenging were not searched out.

It should be noted that primary degradation can lead to the production of substances that are more harmful than the parent compound. This subject is not addressed extensively in this review, although we do report whether the dissipation time includes full mineralization, and any information found on metabolites formed. The concept of biodegradation has little or no meaning for inorganic compounds and metals. Metals will not decompose, but complexation or changes in speciation for them may occur during transport through the different environmental compartments, and thus their intrinsic properties and availability also may be altered. Such potential behavior, however, was beyond the scope of this review.

3.4   Bioaccumulation

The potential of a compound to bioaccumulate is characterized herein, and is expressed by using the bioconcentration factor (BCF). The BCF is the concentration of the chemical in an organism divided by the concentration that exists in the surrounding environment, providing that uptake occurs only through absorption from water via the respiratory surface (e.g., gills). BCFs can only be derived under laboratory conditions, when dietary uptake is minimized, or by theoretical estimation. The $\log K_{OW}$ (octanol–water partition coefficient) value is often used to estimate the BCF and to indicate what the probable bioaccumulation potential for organic compounds is. Generally, there is a good correlation between $\log K_{OW}$ and BCF values (Shüürmann et al. 2007), because compounds having a high $\log K_{OW}$ also have a high tendency to partition to lipids, and therefore possess a high potential for bioaccumulation. The $\log K_{OW}$ value for each HFFR compound addressed is given in the physical–chemical properties section. It should be noted that substances that are rapidly metabolized will have a low bioaccumulation potential, even if they have a high $\log K_{OW}$ (Gobas et al. 2003; Wu et al. 2008). It is currently not possible to include a more refined assessment of bioaccumulation potential, because of the paucity of information that currently exists on the metabolism rates for the HFFRs.

3.5 Toxicity

The in vivo and in vitro toxicity data available for the HFFRs were addressed separately. In vivo toxicity was also addressed separately for ecotoxicity data (from now on merely referred to as ecotoxicity) and mammalian endpoints (often lethal dose studies on rodents).

3.5.1 Ecotoxicity

In vivo aquatic ecotoxicity data are usually reported as lethal concentrations (LC_{50}). In some studies, exposure did not produce an effect. In such cases, the no observed effect concentration (NOEC) is reported, being the highest concentration tested that did not cause an adverse effect compared to the control. This does imply, however, that higher concentrations might show an effect. Alternatively, the lowest observed effect concentration (LOEC) is reported, if available. Ideally, this value implies that lower concentrations will not show an effect and it is best used in combination with a well-defined NOEC. Nevertheless, often LOEC values were reported if the lowest concentration tested showed an effect and no lower concentrations were tested.

3.5.2 In Vivo Toxicity

In vivo toxicity data are reported as the lethal dose (LD_{50}) for feeding or for dermal or inhalation exposure. It should be noted that, in some studies, exposure was not high enough to reach an LD_{50} value. In such cases, the NOEC or LOEC value is reported.

3.5.3   In Vitro Toxicity

For in vitro toxicity we focused on a limited number of well-defined endpoints (as listed below), in which a cell's or organ's function is clearly affected. Mutagenicity and carcinogenicity are addressed as the genotoxic endpoints. The following end-points were addressed for endocrine toxicity: the activation of the Ah-receptor (also called dioxin receptor, DR), the potency to displace thyroxin from its plasma carrier protein transthyretin (TTR), the formation of possibly active metabolites (bioactivation), and the activation of the estradiol receptor (ER) or the androgenic receptor (AR). Finally, we addressed the following neurotoxic endpoints: cytotox-icity, production of reactive oxygen species (ROS), disruption of calcium homeo-stasis and changes in neurotransmitter levels or neurotransmitter receptor activity. Results from the literature are expressed as either EC_{50}, IC_{50} (Inhibition Concentration), LOEC or NOEC values. The literature search revealed that, as for in vivo (eco)toxicity studies, many different in vitro test methods and systems as utilized. In contrast, mutagenic effects were often limited to results of AMES tests (Mortelmans and Zeiger 2000) and are classified as being simply positive or negative.

3.6   Classification

For risk assessment purposes chemicals are often classified according to the differ-ent categories of potential harm that they may cause. In this review, we based our classification on the European Union REACH regulations (European Commission (EC) No 1907/2006 & 1272/2008 (European Union 2006, 2008)). Therefore, where relevant, each of our tables that display intrinsic properties of HFFRs contain a column in which reported values are disclosed as being either "high," "moderate," or "low." Instead of referring to a disappearance "half-life," we prefer to use the term dissipation time, i.e., DT_{50}. Compounds that are classified as being "very per-sistent" and "very bioaccumulative" (vPvB) are based on an existing system that exists in the Regulation of the European Commission (EC No 1907/2006 (European Union 2006)). Atmospheric dissipation times were reported but were not classified. In Table 2, we show the threshold values for each classification level. Complete concentration–response curves were usually absent for in vitro toxicity tests. When toxicity data came from several different studies, it was generally difficult to clas-sify the risk of a compound, according to our preferred classification scheme, i.e., from "no potency" to "very high potency." Therefore, data are presented as "low toxicity" when no effects were observed, "toxic" when effects were observed, and "not enough data to classify" when data were incompatible with the predefined risk assessment criteria or too few details were provided.

Table 2 Classification for persistence, bioaccumulation, and toxicity (PBT)

Classification	Persistence	Bioaccumulation	Toxicity
High	Not "ready biodegradable" or Soil/sediment or sludge DT_{60+} >28 days Or water (pH 7) DT_{70+} >28 days	BCF >500 $\log K_{OW}$ ≥4	LD_{50} ≤1 mg L^{-1} EC_{50} ≤1 mg L^{-1} LC_{50} ≤1 mg L^{-1}
Moderate	–	–	1 mg L^{-1} < LD_{50} ≤ 10 mg L^{-1} 1 mg L^{-1} < EC_{50} ≤ 10 mg L^{-1} 1 mg L^{-1} < LC_{50} ≤ 10 mg L^{-1}
Low	"Ready biodegradable" Or soil/sediment or sludge DT_{60+} ≤28 days Or water (pH 7) DT_{70+} ≤28 days	BCF <500 $\log K_{OW}$ <4	LD_{50} >10 mg L^{-1} EC_{50} >10 mg L^{-1} LC_{50} >10 mg L^{-1}
vPvB	DT_{50} >60 days (marine, fresh, or estuarine water) or DT_{50} >180 days (soil, marine, fresh, or estuarine water sediment) AND a BCF >5,000		

BCF bioconcentration factor, DT_x dissipation time of $x\%$ of the compound, EC_{50} the concentration that causes 50% effect to the test species population, LD_{50} the concentration that causes 50% mortality of the test species population, $Log\ K_{OW}$ logarithmic octanol–water partitioning coefficient

4 Inorganic Flame Retardants and Synergists

In this section, we address the compounds aluminum trihydroxide (ATH), magnesium hydroxide ($Mg(OH)_2$), ammonium polyphosphate (APP), zinc borate, zinc hydroxystannate (ZHS), and zinc stannate (ZS).

4.1 Aluminum Trihydroxide

Aluminum trihydroxide (ATH, CAS nr 21645-51-2) is a weak inorganic acid. It is a hydrate, which means that in its solid form it contains water ($Al(OH)_3 \cdot H_2O$). ATH is commonly used as a smoke suppressor and as a flame retardant synergist, together with other FRs such as organophosphorus compounds (ENFIRO Partners and Leonards (Project Coordinator) 2008). ATH was classified as an HPV chemical in the EU (European Chemicals Bureau 2011). For the USA, the total annual production was given as <450,000 t in 2006 (US EPA 2006).

4.1.1 Physical–Chemical Properties

ATH is solid at environmentally relevant temperatures (−40 to +40°C), since most reported melting points range from 150 to 300°C (European Chemicals Bureau 2000a; Lewis 2000; Martin Marietta Magnesia Specialties LLC (MMMS) et al. 2010).

Our literature search revealed that ATH had a wide range of water solubility, with values ranging from 0.015 to 1.5 mg L^{-1} (1.92E−4 or 1.90E−2 mol m^{-3}) (European Chemicals Bureau 2000a). Two other studies simply refer to ATH as "insoluble" (European Chemicals Bureau 2000a; Rio Tinto Alcan (RTA) 2008a). Clearly, ATH has low water solubility, although its reported solubility values vary by a factor of one hundred. The properties of ATH are listed in Table 3.

4.1.2 Bioaccumulation

In a draft EPA report, it was estimated that the BCF value for ATH is <500 (US EPA 2008), and it was stated in another study that its bioaccumulation potential is low (German Federal Environmental Agency et al. 2001), but neither study gave further details (Table 3).

4.1.3 Toxicity

1. Ecotoxicity
   Reported effect concentrations cover a wide range, and consequently, the classification of ecotoxicity varies between low and high (Table 3). It is not expected that ATH will easily decompose to produce freely dissolved Al^{3+} ions, unless conditions such as a low pH favor Al^{3+} dissociation. The toxicity of aluminum has been extensively discussed elsewhere (Berthon 2002; Kucera et al. 2008) and is not repeated here.
2. In vivo toxicity
   The acute toxicity of ATH to rats is very low, with LD$_{50}$ values higher than 5,000 mg kg^{-1} bwt (Table 3).
3. In vitro toxicity
   Data on ATH were limited (The Subcommittee on Flame-Retardant Chemicals 2000). As shown in Table 3, ATH is not carcinogenic in animal tests (The Subcommittee on Flame-Retardant Chemicals 2000; German Federal Environmental Agency et al. 2001; O'Connell et al. 2004). In one report, it was stated that ATH was mutagenic and cytotoxic, although the concentrations or test conditions were not mentioned (German Federal Environmental et al. 2001). Therefore, the genotoxicity is classified as being low.
   No in vitro endocrine toxicity or neurotoxicity data were reported for ATH. However, it was shown that ATH causes cytostatic activity with induction of neurites at >200 μM in neuroblastoma cells (Zatta et al. 1992). Moreover, at a concentration of >10 μM, ATH did bind to the *N*-methyl-D-aspartate receptor (NMDA-R) in human cerebral cortex (Hubbard et al. 1989). In another study, it was reported that there were adverse effects of ATH on the learning ability in rats and that cholinergic activity was diminished (Bilkei-Gorzo 1993). Despite these adverse neurotoxic effects, there are insufficient data available to classify the overall in vitro toxicity of ATH.

Table 3 Aluminum trihydroxide (ATH, CAS nr 21645-51-2)

	Data	Details	References
Physical–chemical properties			
Molecular weight	78.01		European Chemicals Bureau (2000a)
Melting point	150–220 g mol^{-1} (decomposition)		European Chemicals Bureau (2000a)
Melting point	200°C		MMMS LLC et al. (2010)
Melting point	230°C		Lewis (2000)
Melting point	300°C		Rio Tinto Alcan (RTA) (2008a)
Melting point	2,030°C		European Chemicals Bureau (2000a)
Water solubility	0.015 mg L^{-1}	[at 20°C]	European Chemicals Bureau (2000a)
Water solubility	1.5 mg L^{-1}	[at 20°C]	European Chemicals Bureau (2000a)
Water solubility	Insoluble		European Chemicals Bureau (2000a), RTA (2008a)
Bioaccumulation			
Low	*BCF < 500^{c,e}*	Fish, estimated	US EPA (2008)
Low	Not specified		German Federal Environmental Agency et al. (2001)
Ecotoxicity			
High; aquatic	EC$_{50}$ = 0.8240^e mg L^{-1}	*Daphnia magna*, 48 h	TSCATS, DuPont Central Research 1996 (not found) from draft US EPA (2008)
High; aquatic	EC$_{50}$ = 0.6560^e mg L^{-1}	*Selenastrum capricornutum*, 96 h	TSCATS, DuPont Central Research 1996 (not found) from draft US EPA (2008)
Moderate; aquatic	LC$_{50}$ = 2.6–3.5 mg L^{-1}	Daphnids	Illinois EPA (2007)
Low; aquatic	Not specified	Fish, lethality only at low pH	Illinois EPA (2007)
Low; aquatic	NOEC > 100 mg L^{-1}	Fish (*Salmo trutta*), 96 h; Crustacean (*Daphnia magna*), 48 h; Algae (*Selenastrum capricornutum*), 72 h	European Chemicals Bureau (2000a)
Low; aquatic	Not specified		Stevens and Mann (1999), German Federal Environmental Agency et al. (2001)
In vivo toxicity			
Low	LD$_{50}$ > 5,000 mg kg^{-1} bwt	Rats	European Chemicals Bureau (2000a), Illinois EPA (2007)
Low	Not specified		Stevens and Mann (1999), United Nations Environment Programme (UNEP) (2008)
In vitro toxicity			
Low	Genotoxicity; carcinogenicity	Rats	The Subcommittee on Flame-Retardant Chemicals (2000), O'Connell et al. (2004)

Italic values are predicted: ^{a}Modeled, ^{b}calculated, ^{c}expert judgment
^{e}Not all primary sources are found from (US EPA 2008), also this reference is a draft report, so reported values may be not final

In summary, ATH is a solid at room temperature, and has a low, but uncertain water solubility. Its bioaccumulation potential is estimated to be low and the in vivo and in vitro toxicity of ATH is also low. However, ATH may pose a risk to aquatic communities, with EC_{50} values varying from low to high.

4.2 Magnesium Hydroxide, Mg(OH)$_2$

Magnesium hydroxide (CAS 1309-42-8) is an inorganic salt that consists of hydroxide and magnesium ions. Magnesium hydroxide, $Mg(OH)_2$, used as a flame retardant or flame retardant additive, is very effective in reducing smoke emissions from burning plastics (MMMS LLC et al. 2010). This compound is classified as an HPV chemical in the EU (European Chemicals Bureau 2011). For the USA, total annual production was 45,000 to <227,000 t in 2006 (US EPA 2006).

4.2.1 Physical–Chemical Properties

Magnesium hydroxide is a solid at room temperature, since its melting point is approximately 350°C; oddly, an excessively high melting point of 2,800°C was reported in one study, which seems unlikely (Table 4). This compound has no real boiling point as it will undergo endothermic decomposition at 330 or 340°C, with release of water (AluChem 2003; MMMS LLC et al. 2010). Magnesium hydroxide is insoluble in water (Fisher Scientific 1999 (2008 updated); Albemarle corporation 2003a, b, c; AluChem 2003). An overview of its physical–chemical properties is shown in Table 4.

4.2.2 Bioaccumulation

There are no data available on the bioaccumulation of magnesium hydroxide.

4.2.3 Toxicity

1. Ecotoxicity
   There are no data available on the ecotoxicity of magnesium hydroxide. Magnesium is an essential metal and it is a major component of natural waters (European Chemicals Bureau 2000c). Therefore, it is not expected that this compound has a high aquatic toxicity.
2. In vivo toxicity
   Data on the in vivo toxicity of this compound are quite sparse, with only two acute LD_{50} values (each >5,800 mg kg^{-1} bwt) for rats being reported (Table 4).

Table 4 Magnesium hydroxide ($Mg(OH)_2$, CAS nr 1309-42-8)

	Data	Details	References
Physical–chemical properties			
Molecular weight	58.32 g mol^{-1}		
Melting point	330°C (decomposition)		MMMS LLC et al. (2010)
Melting point	340°C (decomposition)		AluChem (2003)
Melting point	350°C		Fisher Scientific (1999 (2008 updated))
Melting point	2,800°C		Merck & Co. Inc. (2001)
Water solubility	Insoluble mg L^{-1}	[at 25°C]	Fisher Scientific (1999 (2008 updated)), Albemarle corporation 2003a, b, c, AluChem (2003)
In vivo toxicity			
Low	$LD_{50} = 8,500$ mg kg^{-1}	Rats	Merck Chemicals—Product Information (Merck Website)
Low	$LD_{50} = 5,800$ mg kg^{-1} bwt	No details provided	Nabaltec (2009)

Italic values are predicted: [a]Modeled, [b]calculated, [c]expert judgment

3. In vitro toxicity

In vitro toxicity data on magnesium hydroxide ($Mg(OH)_2$) are scarce (Table 4). It is expected that magnesium hydroxide dissociates in the acid environment of the stomach to Mg^{2+}. Therefore, the toxic effects of Mg^{2+} should be included in the risk assessment. For other magnesium salts several toxic effects were described (The Subcommittee on Flame-Retardant Chemicals 2000). There are not enough data to classify the in vitro toxicity of magnesium hydroxide.

In summary, magnesium hydroxide is a solid at room temperature and has low water solubility. Hardly any data are available on the PBT properties of this compound. $Mg(OH)_2$ displayed a low in vivo toxicity in two studies.

4.3 Ammonium Polyphosphate

Ammonium polyphosphate (APP, CAS 68333-79-9) is an ionic inorganic polymeric compound that, due to the polymerization process, consists of a mixture of polymers of different chain lengths and degrees of branching. It is an intumescent flame retardant, which means that the compounds swells when exposed to heat, and thereby reduces heat transfer (ENFIRO Partners and Leonards (Project Coordinator) 2008). In soil and sewage sludge, APP was reported to break down rapidly into ammonia and phosphate (no reported half-life) (German Federal Environmental Agency et al. 2001). When in contact with water APP undergoes slow hydrolysis with the release of ammonium phosphate (Clariant Flame Retardants, pers. comm.).

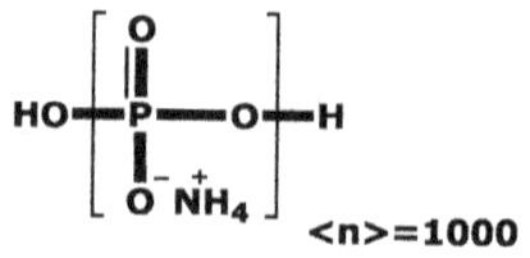

Fig. 1 Schematic representation of the chemical structure of the polymer APP. The product typically consists of a mixture of polymers with an average chain length of 1,000

The indicative production volume for the APP market in Europe is >1,500 t year^{-1} in 1995 (World Health Organization (WHO) 1997). APP is currently classified as an HPV chemical in the EU (European Chemicals Bureau 2011). For the USA, total annual production was given as 45,000 to <227,000 t in 2006 (US EPA 2006).

4.3.1 Physical–Chemical Properties

The physical–chemical properties of polymers strongly depend on the size or length of the polymeric chain. For example, it is usually observed that as chain length increases, melting and boiling temperatures also increase. A common means of expressing the length of a polymer chain is the degree of polymerization, in which the number of monomers incorporated into the chain is quantified. As with other molecules, a polymer's size may also be expressed in terms of molecular weight. Since synthetic polymerization techniques typically yield a polymeric product including a range of molecular weights, the weight is often expressed statistically to describe the distribution of chain lengths present (e.g., average molecular weight). According to a manufacturer (Clariant Flame Retardants, pers. comm.), APP polymers typically have a molecular weight of ca. 100,000 g mol^{-1} (based on an average chain length of 1,000; Fig. 1). We assumed that all measures of physical–chemical properties involve the testing of the technical product (MW ca. 100,000 g mol^{-1}), because to our knowledge, no purified monomeric APP is currently available on the market.

The reported melting point of APP was ≥275°C (European Chemicals Bureau 2000d; German Federal Environmental et al. 2001) and indicates that these polymers are solids at environmentally relevant temperatures. The water solubility of APP is high and was reported as being 10 g L^{-1} or miscible with water (European Chemicals Bureau 2000d; German Federal Environmental et al. 2001). The vapor pressure of this compound is <10 Pa (German Federal Environmental et al. 2001) or <100 Pa (European Chemicals Bureau 2000d). However, these values are misleading (and could wrongly be interpreted as indicating high volatility); we are instead inclined to believe the manufacturer's statement (Clariant Flame Retardants, pers. comm.) that the substance is non-volatile. An overview of the physical–chemical properties of APP is shown in Table 5.

4.3.2 Bioaccumulation

APP has a low bioaccumulation potential (German Federal Environmental et al. 2001; Table 5), although no specific BCF values or other details were given.

Table 5 Ammonium polyphosphate (APP, CAS nr 68333-79-9)

	Data	Details	References
Physical–chemical properties			
Molecular weight	$\approx100,000$ g mol^{-1}		Clariant (pers. comm.)
Melting point	275°C		German Federal Environmental Agency et al. (2001)
Melting point	>275°C		European Chemicals Bureau (2000d)
Melting point	300°C (decomposition)		German Federal Environmental Agency et al. (2001)
Water solubility	10,000 mg L^{-1}	[at 25°C]	European Chemicals Bureau (2000d), German Federal Environmental Agency et al. (2001)
Water solubility	<1,000 mg L^{-1}	[at 25°C]	Budenheim (2010)
Water solubility	<5,000 mg L^{-1}	[at 25°C]	Clariant (2010)
Vapor pressure	<10 Pa	[at 20°C]	German Federal Environmental Agency et al. (2001)
Vapor pressure	<100 Pa	[at 20°C]	European Chemicals Bureau (2000d)
Log K_{ow}	−2.15[a]		ACD/Labs (2011)
Bioaccumulation			
Low	Not specified		German Federal Environmental Agency et al. (2001)
Ecotoxicity			
Low; aquatic	NOEC = 87.6 mg L^{-1}	Algae	UNEP OECD SIDS (2007)
Low; aquatic	EC$_{50}$ = 813–848 mg L^{-1}	Crustacean, *Daphnia magna*, 48 h[d]	McDonald et al. (1996)
Low; aquatic	NOEC > 500 mg/L	Fish, *Danio rerio*, 96 h,	Budenheim (2010)
Low; aquatic	LC$_{50}$ > 500 mg/L	Fish, *Danio rerio*, 96 h	Clariant (2010)
Low; aquatic	LC$_{50}$ = 1,326.0 mg L^{-1}	Fish, *Oncorhynchus mykiss*, 96 h, pH 7	Blahm 1978 (not found) from US EPA (2012)
Low; aquatic	LC$_{50}$ = 123.0 mg L^{-1}	Fish, *Oncorhynchus mykiss*, 96 h, pH 8	Blahm 1978 (not found) from US EPA (2012)
Low; aquatic	LC$_{50}$ > 101 mg L^{-1}	Fish, fresh water	UNEP OECD SIDS (2007)
Low; aquatic	Not specified	NOEC exceeds solubility, low acute aquatic toxicity	European Chemicals Bureau (2000d)
Low; aquatic	LD$_{50}$ > 500 mg L^{-1}	Fish, fresh water	European Chemicals Bureau (2000d)
Moderate; aquatic	EC$_{50}$ = 1.790 mg L^{-1}	Crustaceans (*Daphnia carinata*), 72 h	UNEP OECD SIDS (2007)
Moderate; aquatic	IC$_{50}$ = 10 mg L^{-1}	Algae, *Selenastrum capricornutum*, 96 h[d]	McDonald et al. (1996)
In vivo toxicity			
Low	LD$_{50}$ > 2,000 mg kg^{-1} bwt	Rats	UNEP OECD SIDS (2007)
Low	LD$_{50}$ > 4,740 mg kg^{-1} bwt	Rats	European Chemicals Bureau (2000d)
In vitro toxicity			
Low	Genotoxicity; mutagenicity	*Salmonella* and *E. coli*, AMES test	European Chemicals Bureau (2000d)

Italic values are predicted: [a]Modeled, [b]calculated, [c]expert judgment
[d]APP has the technical name Fire-Trol LCG-R (McDonald et al. 1996)

We estimated the log K_{ow} as −2.15 (ACD/Labs 2011), showing the compound to have high hydrophilicity. Because of the high aqueous solubility of the polymer and the large molecular size, APP is not expected to bioaccumulate (MW ca. 100,000 g mol⁻¹) (Dimitrov et al. 2002).

4.3.3 Toxicity

1. Ecotoxicity
   Data from several aquatic toxicity studies indicate low to moderate toxicity to several algal species, crustaceans, and fish (Table 5).
2. In vivo toxicity
   Two studies on APP were reported in which the toxicity to rats was low (European Chemicals 2000d; UNEP OECD SIDS 2007; Table 5).
3. In vitro toxicity
   Limited in vitro data are available for APP. No carcinogenic, endocrine, or neurotoxic data were found, although AMES test results showed no response for mutagenic activity (Table 5) (European Chemicals Bureau 2000d). This suggests that the chemical has a low genotoxicity. APP is probably hydrolyzed by stomach acids into phosphate and ammonium ions, and various effects could be expected based on the structural similarities with other compounds, e.g., inositol polyphosphates or adenosine polyphosphates. Therefore, there were insufficient data to classify either endocrine- or neurotoxicity. Data on developmental toxicity were not available.

In summary, APP is a solid that has high water solubility and is expected to degrade in natural environments. It predominantly exerts a low toxicity on the aquatic community, although two authors reported moderate toxicity to daphnids or algae. Toxicity to rats is low, as is the reported in vitro toxicity, although the number of available studies is limited.

4.4 Zinc Borate (ZB)

Zinc borates (ZBs) exist in different mineral compositions and have different zinc oxide and borate ratios. Additionally, some borates contain structurally bound water (hydrates). The specific compound discussed in this review is $2ZnO \cdot 3B_2O_3$ (CAS 138265-88-0 or 12767-90-7), a non-hydrate. It is being used as a flame retardant synergist and smoke suppressor (European Flame Retardants Association (EFRA) and Cefic 2006). ZB breaks down to zinc hydroxide and boric acid under natural conditions (EFRA and Cefic 2006). ZB is currently classified as an LPV chemical in the EU (European Chemicals Bureau 2011). No information is available on production volumes in the USA.

4.4.1 Physical–Chemical Properties

There is very little information on the physical–chemical properties of ZB, or even for other mineral compositions of this compound. It is solid at room temperature and decomposes at 650°C (synthetic ZB) (Borax 2004). ZB has a solubility in water of 2.8 g L^{-1} (6.44 mol m^{-3}) (Borax 2004).

4.4.2 Bioaccumulation

No information on bioaccumulation of ZB is available.

4.4.3 Toxicity

Zinc is an essential element for animals (Maret and Sandstead 2006) and plants (EFRA and Cefic 2006). However, intake of more than 100–300 mg zinc per day results in adverse health effects (Fosmire 1990; Rout and Das 2003). Depending on the prevailing conditions in the environment, ZB can decompose to produce freely dissolved zinc ions. The toxicity of zinc has been studied extensively (Barceloux 1999; Cummings and Kovacic 2009; Nagajyoti et al. 2010) and will not be repeated here. The same holds true for the toxicity of boric acid (European Chemicals Bureau 2000b).

1. Ecotoxicity
   According to the few reports available, zinc borate has a high aquatic toxicity to daphnids, algae, and several fish species (Table 6).
2. In vivo toxicity
   There are only a few studies on ZB toxicity (Table 6). An LOEC value of 0.91 mg L^{-1} day^{-1} was reported for humans.
3. In vitro toxicity
   Limited information is available in the literature on the in vitro toxicity of ZB. There are no data on carcinogenicity, endocrine disruption, or neurotoxicity. Zinc borate is not mutagenic (Illinois EPA 2007; Table 6). As ZB probably readily breaks down in the stomach to zinc oxide (ZnO) and boric acid (H$_3$BO$_3$), these compounds should also be included in any risk assessment. Although there are extensive databases on the toxicity of zinc oxide and boric acid (The Subcommittee on Flame-Retardant Chemicals 2000), there are not enough data to classify the in vitro toxicity of ZB.

In summary, zinc borate is a solid with moderate aqueous solubility. It has high aquatic toxicity, whereas reported values for in vivo toxicity vary from low to high. According to the results of one study, zinc borate has low mutagenicity; however, there is a lack of information on the in vitro toxicity of the compound.

Table 6 Zinc borate (ZB, CAS nr 138265-88-0 or 12767-90-7)

	Data	Details	References
Physical–chemical properties			
Molecular weight	434.66 g mol^{-1}		
Melting point	650°C (decomposition)		Borax (2004)
Water solubility	2,800 mg l^{-1}	[at 25°C]	Borax (2004)
Ecotoxicity			
High; aquatic	EC_{50}=0.015–0.178 mg L^{-1}	Algal inhibition	Illinois EPA (2007)
High; aquatic	EC_{50}=0.068–1.59 mg L^{-1}	Daphnia	Illinois EPA (2007)
High; aquatic	LC_{50}=0.59–5.9 mg L^{-1}	Fish	Illinois EPA (2007)
High; aquatic	Not specified	Aquatic species	European Flame Retardants Association (EFRA) and Cefic (2006), UNEP (2008)
In vivo toxicity			
Low	LD_{50}>10,000 mg kg^{-1}	Rats and rabbits, oral and dermal exposure	EFRA and Cefic (2006)
Low	LD_{50}>2,000 mg kg^{-1}	Rat, mice, dog	Illinois EPA (2007)
High	LOEC=0.91 mg L^{-1} day^{-1}	Humans, zinc blood effects	Illinois EPA (2007)
High	Can be harmful to the unborn	Not specified	McPherson et al. (2004)
In vitro toxicity			
Low	Genotoxicity; mutagenicity	–	Illinois EPA (2007)

–, no effects observed

4.5   Zinc Hydroxystannate

Zinc hydroxystannate (ZHS; ZnSn(OH)$_6$, CAS 12027-96-2) is an inorganic, bimetallic hydroxide used as a smoke suppressant (William Blythe 2010a). No information is available on production volumes in the EU (European Chemicals Bureau 2011). For the USA, total annual production was <227 t in 2006 (US EPA 2006).

4.5.1   Physical–Chemical Properties

ZHS decomposes at 180–200°C (Australian Government Regulator of Industrial Chemicals 1994; RTA 2008b; ITRI 2009), at these temperatures dehydroxylation occurs, releasing water from the crystal (William Blythe Ltd., pers. comm.). It has a low water solubility of 1 mg L^{-1} (0.0035 mol m^{-3}) (Australian Government Regulator of Industrial Chemicals 1994; RTA 2008b) (primary source not stated). The vapor pressure and log K_{OW} of ZHS were reported to be low (Table 7).

Table 7 Zinc hydroxystannate (ZHS, CAS nr 12027-96-2)

	Data	Details	References
Physical–chemical properties			
Molecular weight	286.11 g mol⁻¹		ITRI (2009)
Melting point	>180°C (decomposition)		Australian Government Regulator of Industrial Chemicals (1994), RTA (2008b), William Blythe Ltd. (pers. comm.)
Melting point	200°C (decomposition)		Australian Government Regulator of Industrial Chemicals (1994), RTA (2008b)
Water solubility	1 mg L⁻¹	[mg L⁻¹ at 20°C]	Australian Government Regulator of Industrial Chemicals (1994), William Blythe (2010a)
Water solubility	Insoluble		William Blythe (2010a)
Vapor pressure	<10 Pa	[at 20°C]	Australian Government Regulator of Industrial Chemicals (1994)
Log K_{ow}	<−1.05		Australian Government Regulator of Industrial Chemicals (1994)
Log K_{ow}	<−1		William Blythe Ltd. (pers. comm.)
Log K_{ow}	<0.09		RTA (2008b)
Bioavailability			
Low	*Low potential estimated*[c]	Based on low K_{ow} value, no specified data	William Blythe Ltd. (pers. comm.)
Ecotoxicity			
Low; aquatic	LD_{50} >3.3 mg L⁻¹	Fish, NOEC>water solubility	Joseph Storey & Co. Ltd. (1994)
Low; aquatic	EC_{50} >3.3 mg L⁻¹	Crustaceans, NOEC>water solubility	Joseph Storey & Co. Ltd. (1994)
Low; aquatic	LC_{50} >0.079 mg L⁻¹	Rainbow Trout, LC_{50} >water solubility, acute, no details provided	William Blythe (2010a)
Low; aquatic	E_{50} >0.023 mg L⁻¹	*Daphnia magna*, EC_{50} >water solubility, 48 h, no details provided	William Blythe (2010a)
In vivo toxicity			
Low	LD_{50} >5,000 mg kg⁻¹ bwt	Rats	Gardner (1988a), Joseph Storey & Co. Ltd. (1994)
Low	LD_{50} >2,466 mg kg⁻¹	Rats, dermal exposure	Joseph Storey & Co. Ltd. (1994)
Low to moderate	LD_{50} >4.3 mg L⁻¹	Rats, inhalation exposure	Joseph Storey & Co. Ltd. (1994)
In vitro toxicity			
Low	Genotoxicity; mutagenicity	−, Salmonella, with and without metabolic activation (S9), AMES test	Australian Government Regulator of Industrial Chemicals (1994), William Blythe Ltd. (pers. comm.)

Italic values are predicted: [a]Modeled, [b]calculated, [c]expert judgment

−, no effects observed

4.5.2   Bioaccumulation

According to one producer of ZHS, the bioaccumulation potential is estimated to be low, based on its low water solubility and low K_{ow} value (Table 7). There is no further information on the bioaccumulation of ZHS.

4.5.3   Toxicity

As stated previously for ZB, zinc is an essential element for animals (Maret and Sandstead 2006) and plants (EFRA and Cefic 2006). However, intake of more than 100–300 mg zinc per day produces adverse health effects (Fosmire 1990; Rout and Das 2003). Zinc hydroxide can decompose to produce freely dissolved Zn^{2+} ions, depending on the prevailing conditions in the environment. The toxicity of zinc has been discussed extensively elsewhere (Barceloux 1999; Cummings and Kovacic 2009; Nagajyoti et al. 2010) and is not repeated here.

1. Ecotoxicity
   Low ecotoxicity of ZHS for fish and crustaceans was reported in two studies, in which the NOECs and EC_{50s} exceeded the water solubility (Joseph Storey & Co. Ltd. 1994; William Blythe 2010a). In the latter study, the EC_{50} and LC_{50} values exceeded 0.02 mg L^{-1}. However, this is still a very low concentration and the NOEC or LOEC values from this study are inconclusive.
2. In vivo toxicity
   There is very limited information on the in vivo toxicity of ZHS. A few studies provided data to show low acute toxicity to orally exposed rats (Table 7). Low to moderate toxicity was reported for ZHS in an inhalation exposure study (Joseph Storey & Co. Ltd. 1994). However, because this LD_{50} value was reported as being "greater than" a moderate value, it is not clear whether this value represented just the highest concentration tested, or whether there was an effect observed at this level.
3. In vitro toxicity
   In vitro toxicity data for ZHS are also scarce, and no carcinogenic, endocrine disrupting, or neurotoxic data were available. One author reported no mutagenic activity in an AMES test performed with and without metabolic activation (Australian Government Regulator of Industrial Chemicals 1994).

In summary, ZHS is a solid that has low water solubility. It presumably has a low aquatic toxicity. With only two studies available, the effects on rats vary from low to moderate. There are not enough data to fully classify the in vitro toxicity of the compound; only one study existed, and no mutagenic effects in an AMES test was reported therein.

Table 8 Zinc stannate (ZS, CAS nr 12036-37-2)

	Data	Details	References
Physical–chemical properties			
Molecular weight	232.10 g mol^{-1}		
Melting point	>397°C (decomposition)		RTA (2008b)
Melting point	>570°C		Gelest (2008), ITRI (2009)
Water solubility	Insoluble		William Blythe (2010b)
Water solubility	1 mg L^{-1}	[at 20°C]	Gelest (2008)
Water solubility	13 mg L^{-1}	[at 25°C]	Rio Tinto Alcan (RTA) (2008c)
Vapor pressure	<0.13 Pa	[at 25°C]	Gelest (2008)
Vapor pressure	<10 Pa	[at 20°C]	William Blythe (2010b)
Bioavailability			
Low	*Low potential estimated^c*	Based on low K_{OW} value, no specified data	William Blythe Ltd. (pers. comm.)
Ecotoxicity			
Low; aquatic	LC$_{50}$>0.079 mg L^{-1}	Rainbow Trout, LC$_{50}$>water solubility, acute, no details provided	William Blythe (2010b)
Low; aquatic	E_{50}>0.023 mg L^{-1}	*Daphnia magna*, EC$_{50}$>water solubility, 48 h, no details provided	William Blythe (2010b)
In vivo toxicity			
Low	LD$_{50}$>5 g kg^{-1} bwt	Rats	Gardner (1988b)

Italic values are predicted: ^{a}Modeled, ^{b}Calculated, ^{c}Expert judgement

4.6 Zinc Stannate

Zinc stannate (ZS; $ZnSnO_3$, CAS 12036-37-2) is an inorganic, bimetallic oxide used as a smoke suppressant (William Blythe 2010b). No information is available on production volumes in the EU (European Chemicals Bureau 2011). For the USA, total annual production was <227 t in 2006 (US EPA 2006).

4.6.1 Physical–Chemical Properties

ZS decomposes at 397–570°C (Gelest 2008; RTA 2008c; ITRI 2009) (Table 8). It has low solubility in water that varies between 1 and 13 mg L^{-1} (0.0043 and 0.056 mol m^{-3}) (Gelest 2008; RTA 2008c) (primary sources are not stated).

4.6.2 Bioaccumulation

According to one producer, ZS is estimated to have a low bioaccumulation potential, based on low water solubility and low K_{ow} values (Table 7). There is no further information on the bioaccumulation of ZS.

4.6.3 Toxicity

The toxicity of the zinc ion is discussed in the section of ZB and ZHS and is not repeated here. The toxicity of ZS is discussed below.

1. Ecotoxicity
   Low ecotoxicity of ZS for fish and crustaceans was reported in one study (Table 8), in which LC_{50} and EC_{50} values probably exceeded the water solubility (William Blythe 2010b). In this study, the EC_{50} and LC_{50} values were >0.02 mg L^{-1}. However, this is still a very low concentration and it is not clear what the definitive NOEC or LOEC values from this study were.
2. In vivo toxicity
   There is very limited information about the in vivo toxicity of ZS. A low acute toxicity to orally exposed rats was reported in one study (Gardner 1988b; Table 8).
3. In vitro toxicity
   There is no information on carcinogenic, endocrine disrupting, or neurotoxic effects of ZS.

In summary, zinc stannate is a solid with low water solubility. One author reported low toxic effect on rats. There is an obvious lack of data on other PB&T properties.

5   Organophosphorus Flame Retardant Compounds and Their Salts

The following organophosphorus HFFRs are discussed in this section: triphenylphosphate (TPP), resorcinol bis(diphenylphosphate) (RDP), bisphenol-A bis(diphenylphosphate) (BDP), dihydrooxaphosphaphenanthrene (DOPO), and aluminum diethylphosphinate (ALPI).

5.1   Triphenylphosphate

Triphenylphosphate (TPP, CAS 115-86-6) is an aryl phosphate, mainly being used as a flame retardant in polymers (European Chemicals Bureau 2002), and is the best studied compound of the selected HFFRs (Hoenicke et al. 2007; Bergh et al. 2011). It is present in all environmental compartments, ranging from, e.g., air (23.2 ng m^{-3}) (Danish EPA et al. 1999) to fish (21–180 ng g^{-1} (Sundkvist et al. 2010)). The global production (excluding East Europe) was estimated to be 20,000–30,000 t in one study (UNEP OECD SIDS 2002a). Of this production estimate, approximately 25% was produced in Western Europe, 40% in the USA and 35% in Asia by 15 producers (UNEP OECD SIDS 2002a). TPP is classified as an HPV chemical in the EU (European Chemicals Bureau 2011). For the USA, the total annual production was given as 4,500 to <22,700 t in 2006 (US EPA 2006).

5.1.1 Physical–Chemical Properties

TPP is solid at environmentally relevant temperatures and has a relatively low melting point of approximately 50°C (European Chemicals Bureau 2000f; Hilal et al. 2003; Merck & Co. Inc. 2006; US EPA 2011). Its water solubility is low, showing a wide range from 55 μg/L to 5 mg L^{-1} (1.70E–4 to 1.43E–2 mol m^{-3}) and thereby varies by a factor of about 100 (Saeger et al. 1979; European Chemicals Bureau 2000f; Hilal et al. 2004; Eckert and Klamt 2010; US EPA 2011). In Table 9, we present an overview of the physical–chemical properties of TPP. The low water solubility, Henry's law constant, and K_{AW} vs. the high log K_{OW} indicate that, once released into the environment, TPP probably partitions mainly into organic and lipid-rich compartments such as soil and biota.

5.1.2 Persistence

Degradation data on TPP in the atmosphere, water and soil/sediment, are also presented in Table 9. Atmospheric half-lives are predominantly determined by photolysis (breakdown of the compound by light), and therefore the half-life of TPP is often measured by testing the photolysis rate. In water, degradation can occur by abiotic, e.g., hydrolysis (reaction of the compound with water), or biotic (mediated by microorganisms) mechanisms.

The persistence of this compound is classified as varying from high to low (Table 9). The fastest atmospheric degradation rate of TPP was reported to be a few hours (European Chemicals Bureau 2000f), whereas the longest degradation time was reported to be 406 days in water (US EPA 2005; European Chemicals Bureau et al. 2007).

Given the large variation in the persistence data, we additionally estimated TPP degradation by using EPI Suite 4.1. The resulting degradation half-lives from these estimates were 24 h in air, 900 h in water, and 1,800 h in soil; these estimated values are within the ranges of experimental values. On the basis of these data, this substance is expected to be stable and persistent in the environment.

Table 9 Triphenylphosphate (TPP, CAS nr 115-86-6)

	Data	Details	References
Physical–chemical properties			
Molecular weight	326.29 g mol^{-1}		ICL (pers. comm.)
Melting point	48–50°C		European Chemicals Bureau (2000f), Merck & Co. Inc. (2006)
Melting point	49–50°C		US EPA (2011)
Melting point	50.5°C		Hilal et al. (2003), US EPA (2011)
Melting point	86.5 ^{a}°C		Hilal et al. (2004)
Water solubility	5.55E–2^a mg L^{-1}	[at 25°C]	European Chemicals Bureau (2000f)
Water solubility	2.5E–2 mg L^{-1}	[at 25°C]	European Chemicals Bureau (2000f)
Water solubility	7.5E–1 mg L^{-1}	[at 25°C]	ICL (pers. comm.)
Water solubility	<1 mg L^{-1}	[at 25°C]	US EPA (2011)
Water solubility	1.03^a mg L^{-1}	[at 25°C]	Saeger et al. (1979), European Chemicals Bureau (2000f)
Water solubility	1.9 mg L^{-1}	[at 25°C]	Eckert and Klamt (2010)
Water solubility	2.66^a mg L^{-1}	[at 25°C]	US EPA (2011)
Water solubility	4.67^a mg L^{-1}	[at 25°C]	European Chemicals Bureau (2000f)
Vapor pressure	1.50E–6 Pa	[at 25°C]	Eckert and Klamt (2010)
Vapor pressure	4E–6^a Pa	[at 25°C]	US EPA (2011)
Vapor pressure	6.29E–5^a Pa	[at 25°C]	ACD/Labs (2011)
Vapor pressure	1.65E–4^a Pa	[at 25°C]	ICL (pers. comm.)
Vapor pressure	8.35E–4 Pa	[at 25°C]	Dobry and Keller (1957)
Vapor pressure	8.37E–4 Pa	[at 25°C]	Hilal et al. (2003)
Vapor pressure	3.82E–2^a Pa	[at 25°C]	US EPA (2011)
Henry's law constant	4.03E–3^a Pa [m^3 mol^{-1}]		US EPA (2011)
Henry's law constant	3.35E–1 Pa [m^3 mol^{-1}]		Syracuse Research Corporation (SRC) (2006)
Henry's law constant	1.22 Pa [m^3 mol^{-1}]		SRC (2006)
Log K_{ow}	4.59		Saeger et al. (1979)
Log K_{ow}	4.60		European Chemicals Bureau et al. (2007)
Log K_{ow}	<4.77		

Log K_{OW}	4.9^a		Eckert and Klamt (2010)
Log K_{OW}	6.78^a		Hilal et al. (2004)
Log K_{AW}	-3.87		US EPA (2011)
Log K_{AW}	-6.66^a		Eckert and Klamt (2010)
Persistence			
Low	Not specified; primary source not found		Clean Production Action et al. (2007)
Low; water	$DT_{50} = 1.2–2$ days	pH 8.8, natural water	World Health Organization (WHO) (1991), UNEP OECD SIDS (2002b)
Low; water	$DT_{50} = 1.3$ days	21°C, pH 9.5, hydrolysis	Howard and Deo (1979) from UNEP OECD SIDS (2002b)
Low; water	$DT_{50} = 3$ days	pH 9, 25°C	European Chemicals Bureau (2000f)
Low; water	$DT_{50} = 3$ days	pH 9, hydrolysis	Mayer et al. (1981)
Low; water	$DT_{100} <7$ days	River water	WHO (1991)
Low; water	$DT_{50-100} <8$ days	River die-away test (presumably primary degradation)	US EPA (2005)
Low; water	$DT_{50} <5$ days	Hydrolysis, 20°C, pH 9	US EPA (2005), European Chemicals Bureau et al. (2007)
Low; water	$DT_{50} = 7.5$ days	21°C, pH 8.2, hydrolysis	Howard and Deo (1979), UNEP OECD SIDS (2002b)
Low; water	$DT_{50} = 19$ days	pH 7, hydrolysis	Mayer et al. (1981)
Low; water	$DT_{50} = 19$ days	pH 7, 25°C	European Chemicals Bureau (2000f)
Low; soil and sediment	$DT_{50} = 3–12$ days	(An)aerobic, river water/sediment and pond sediment	European Chemicals Bureau et al. (2007)
Low; soil and sediment	$DT_{50} = 21$ days	Anaerobic	UNEP OECD SIDS (2002b)
Low; sludge	$DT_{83-94} <28$ days (ready biodegradable)		UNEP OECD SIDS (2002b), US EPA (2005)
Low; sludge	Inherently biodegradable, degrades rapidly in pond and river sediment (not specified)	Aerobic	Danish EPA et al. (2007)

(continued)

Table 9 (continued)

	Data	Details	References
High; water	$DT_{50} > 28$ days	pH 5, 25°C	European Chemicals Bureau (2000f)
High; water	$DT_{50} > 28$ days	pH 5, hydrolysis	Mayer et al. (1981)
High; water	$DT_{50} = 37.5^a$ days (900 h)	Primary degradation	US EPA (2011)
High; water	$DT_{50} = 366$ days	Hydrolysis, 20°C, pH 3	US EPA (2005), European Chemicals Bureau et al. (2007)
High; water	$DT_{50} = 406$ days	Hydrolysis, 20°C, pH 7	US EPA (2005)
High; soil and sediment	$DT_{50} = 32$ days	Anaerobic; 5 mg kg^{-1}, 20°C, BBA standard soil 2.2 1993, loamy sand	European Chemicals Bureau (2000f)
High; soil and sediment	$DT_{50} = 37$ days	Aerobic	European Chemicals Bureau (2000f), UNEP OECD SIDS (2002b)
High; soil and sediment	$DT_{22} = 40$ days	Aerobic mineralization, 90% primary degradation, river sediment	UNEP OECD SIDS (2002b)
High; soil and sediment	$DT_{10} = 40$ days (river sediment)	Anaerobic	US EPA (2005)
High; soil and sediment	$DT_{50} = 50$–60 days (pond hydrosoil)	(An)aerobic	Muir et al. (1989), European Chemicals Bureau et al. (2007)
High; soil and sediment	$DT_{50} = 75^a$ days (1,800 h)	Soil, primary degradation	US EPA (2011)
High; soil and sediment	$DT_{50} = 377.5^a$ days (8,100 h)	Sediment, primary degradation	US EPA (2011)
High; soil and sediment	Partially degradable in river sediment and soil (not specified)	Anaerobic	Danish EPA et al. (2007)
n.c.; atmospheric	$DT_{50} < 1$–12 h		European Chemicals Bureau (2000f)
n.c.; atmospheric	$DT_{50} = 12$ h	Photolysis	US EPA (2005), European Chemicals Bureau et al. (2007)
n.c.; atmospheric	$DT_{50} = 12$ h	Primary degradation	European Chemicals Bureau et al. (2007)
n.c.; atmospheric	$DT_{50} = 23.7^a$ h	Primary degradation, modeled	(US EPA 2011)
n.c.; atmospheric	$DT_{50} = 36^e$ h	Photolysis	Draft UK Environment Agency et al. (2009b)

Bioaccumulation

Low	BCF=110–144	Fish species, fresh water	UNEP OECD SIDS (2002b)
	BCF<50	*Lemna minor* (duck weed) and *Typha* sp. (cat tail)	
Low	BCF=420[e] kg L^{-1}	Fish	Draft UK Environment Agency et al. (2009b)
Low	BCF=73.18[g] and	Not specified	CEPA (2007)
	BCF=74.23[a] kg L^{-1} wet-wt		
Low to high	BCF=0.06–271 (*1,800[b] calculated*)	Fish species, fresh water	Danish EPA et al. (2000), European Chemicals Bureau (2000f), McPherson et al. (2004)
Low to high	BCF=110–500	Fish species, *Carassius auratus* and *Oryzias latipes*	Sasaki et al. (1981)
Low to high	BCF=6–18,900	Fish species	Value range from WHO (1991)[d]
Low to high	BCF=18–2,590	Species not specified	Illinois EPA (2007)
Low to high	BCF=132–1,743	Fish; 132–264 rainbow trout and 218–1,743 fathead minnow	US EPA (2005)

Ecotoxicity; aquatic

Low to high	LC$_{50}$=0.24–290 mg L^{-1}	Various fresh water species	European Chemicals Bureau (2000f)
Low to high	LC$_{50}$=0.36–290 mg L^{-1}	Fish	WHO (1991), Illinois EPA (2007)
Moderate	EC$_{50}$=5 mg L^{-1}	Algae	WHO (1991)
Moderate	EC$_{50}$=2.0 mg L^{-1}	Algae	US EPA (2005)
Moderate	EC$_{50}$=2 mg L^{-1}	Algae, growth inhibition	Danish EPA et al. (2007)
Moderate	LC$_{50}$=1.2 mg L^{-1}	Daphnids	US EPA (2005)
Moderate	LC$_{50}$=1.0–1.2 mg L^{-1}	Daphnids	Illinois EPA (2007)
Moderate	LC$_{50}$=1.0 mg L^{-1}	Daphnids	WHO (1991)
Moderate to high	LC$_{50}$=1–1.35 mg L^{-1}	Crustaceans	Danish EPA et al. (2007)
Moderate to high	LC$_{50}$=0.36–1.2 mg L^{-1}	Fish	Danish EPA et al. (2007)
Moderate to high	LC$_{50}$>0.32–1.2 mg L^{-1}	Various fresh water species, with solvent	European Chemicals Bureau (2000f)
Moderate to high	LD$_{50}$=0.290–290 mg L^{-1}	Fish	Danish EPA et al. (2000)

(continued)

Table 9 (continued)

	Data	Details	References
Moderate to high	$EC_{50} = 0.26$–2.0 mg L^{-1}	Algae, growth inhibition	Danish EPA et al. (2000), Illinois EPA (2007)
Moderate to high	$EC_{50} > 0.18$–1.00 mg L^{-1}	Invertebrates	European Chemicals Bureau (2000f); UNEP OECD SIDS (2002b)
High	$LC_{50} = 0.870$ mg L^{-1}	Fish	US EPA (2005)
High	$LC_{50} = 0.4$ mg L^{-1}	Fish	Mayer et al. (1981)
High	$LC_{50} = 0.31$ mg L^{-1}	Trout	CEPA (2007)
High	$EC_{50} = 0.26$–0.5 mg L^{-1}	Algae, growth inhibition	WHO (1991), European Chemicals Bureau (2000f)
High	$LC_{50} = 0.140$–0.600 mg L^{-1}	Algae, growth inhibition, chronic	US EPA (2005)
High	$LC_{50} = 0.1$ mg L^{-1}	Daphnids, chronic exposure	US EPA (2005)
High	$LC_{50} = 0.09$–0.140 mg L^{-1}	Fish, chronic exposure	US EPA (2005)
High	Not specified (primary source not found)		Clean Production Action et al. (2007), European Chemicals Bureau et al. (2007)
n.c.	$NOEC = 0.0014$ mg L^{-1}	Fish, chronic exposure	WHO (1991), Illinois EPA (2007)
n.c.	$EC_{10} = 0.037$ mg L^{-1}	Fish (*Oncorhynchus mykissi*), 30 days, since EC_{10} is very low, might expect high ecotoxicity	ICL (pers. comm.)
n.c.	Lowest chronic NOEC <0.01 mg L^{-1}		UK Environment Agency et al. (2003)
n.c.	$NOEC = 0.1$ mg L^{-1}	Algae, growth inhibition, chronic exposure	Danish EPA et al. (2007)
n.c.	$NOEC = 0.1$ mg L^{-1}	Daphnids, chronic exposure	Illinois EPA (2007)
n.c.	$NOEC = 0.25$–2.5 mg L^{-1}	Algae (*Selenastrum capricornutum, Scenedesmus subspicatus, Chlorella vulgaris*), 96 h	ICL (pers. comm.)
In vivo toxicity			
Low	$LD_{50} = 1{,}300$ mg kg^{-1} bwt	Mice	Danish EPA et al. (2007)
Low	$LD_{50} = 1{,}300$–$10{,}800$ mg kg^{-1} bwt	Mice, rats, cats	WHO (1991), European Chemicals Bureau (2000f), UNEP OECD SIDS (2002b)
Low	$LD_{50} > 3{,}000$ mg kg^{-1} bwt	Mice, rats, cats	Danish EPA et al. (2007)

Low	LD_{50} >3,000 mg kg⁻¹ bwt	Guinea pigs, hen	WHO (1991), European Chemicals Bureau (2000f), UNEP OECD SIDS (2002b)
Low	LD_{50} = 3,500–20,000 mg kg⁻¹ bwt	Rats	Merck Chemicals—Product Information (Merck Website), Danish EPA et al. (2007), European Chemicals Bureau et al. (2007), Illinois EPA (2007)
Low	LD_{50} >5,000 mg kg⁻¹ bwt	Mice, rats, rabbits	US EPA (2005)
Low	LD_{50} >7,900 mg kg⁻¹	Rabbit, dermal exposure	US EPA (2005)
Low	LD_{50} >7,900 mg kg⁻¹	Rabbit, dermal exposure	Merck Chemicals—Product Information (Merck Website), Danish EPA et al. (2007)
Low	LD_{50} >8,000 mg kg⁻¹	Mammal (not specified), dermal exposure	US EPA (2005)
In vitro toxicity			
Low	Genotoxicity; mutagenicity	Salmonella, AMES test	Zeiger et al. (1988), Danish EPA et al. (1999), Illinois EPA (2007)
Low	Genotoxicity; carcinogenicity	Expected low for human and animals	Washington State Department of Ecology and Department of Health (2006), US Department of Health and Human Services et al. (2009), latter is draft[e]
Low	Neurotoxicity; cytotoxicity	IC_{50} = 800 µM PC12 cells	Flaskos et al. (1994)
High	Neurotoxicity; cholinesterase inhibitor not further specified		Bingham et al. (2001)
n.c.	Endocrine toxicity; metabolization	Metabolites formed: diphenylphosphate (rat liver microsomes) and diphenyl p-hydroxyphenol phosphate (houseflies, rats, and goldfish)	Eto et al. (1975), Sasaki et al. (1981), Snyder (1990)

Italic values are predicted: [a]Modeled, [b]calculated, [c]expert judgment

n.c., not enough data to classify

[d]For BCFs we could not find all primary references from/or reported values were not corresponding to the references stated in (WHO 1991)

[e]These references (UK Environment Agency et al. 2009b; US Department of Health and Human Services et al. 2009) are draft reports, so reported values may be not final

5.1.3   Bioaccumulation

The bioaccumulation of TPP has mainly been studied in fish, and results vary from low to high for different species (Table 9). A BCF as low as 0.06 was reported for *Phoxinus phoxinus*, a fresh water minnow (Bengtsson et al. 1986) (European Chemicals Bureau 2000f), and a BCF as high as 1,743 for *Pimephales promelas*, the fathead minnow (US EPA 2005). Our literature research revealed remarkably diverging opinions on what is considered to be a high bioconcentration factor. The Clean Production Action reported a high BCF for TPP of >100 (Clean Production Action et al. 2007), whereas the Illinois EPA reported a low potential, with a BCF as high as 2,590 (Illinois EPA 2007). In comparison, the REACH criterion states that a compound with a BCF larger than 500 is classified as bioaccumulative (European Union 2008; Table 2). We believe that the order of magnitude in the REACH guidelines (i.e., BCF >500) is more realistic concerning what should be considered as potentially bioaccumulative. On the basis of this classification, most experimental data indicate that this compound is bioaccumulative.

5.1.4   Toxicity

1. Ecotoxicity
   The aquatic toxicity of TPP is, in many cases, high to many types of fish, algae, and crustaceans, since the LC_{50s} recorded for these species is about or lower than 1 mg L^{-1} (Table 9).
2. In vivo toxicity
   Many low effect concentrations were reported for higher organisms (e.g., rodents), as shown in Table 10. Hence, the toxicity to these species is considered to be low.
3. In vitro toxicity
   An overview of the in vitro toxicity data is shown in Table 9. TPP is not considered to be a potent anticholinesterase agent, however, exposure to 150 to 300 mg kg^{-1} bwt^{-1} TPP does inhibit cholinesterase in rats, in vitro as well as in vivo (Bingham et al. 2001). TPP may be metabolized into diphenyl-hydroxyphenolphosphate and diphenylphosphate (Eto et al. 1975; Snyder 1990). Several in vitro effects were reported in rats exposed to a commercial cresyldiphenylphosphate product that contains TPP (Vainiotalo et al. 1987). Several neurotoxic effects were observed in in vitro studies, e.g., cytotoxicity in PC12 cells and inhibition of the GABA-regulated chloride channel (Gant et al. 1987; Padilla et al. 1987; Vainiotalo et al. 1987; Flaskos et al. 1994). However, the neurotoxicity of TPP has been debated since the early studies of Smith et al. (1930, 1932), because neurotoxic changes in animals after short-term exposure were not identified in other studies (Wills et al. 1979). Since TPP exposure does result in several toxic effects, the in vitro endocrine and neurotoxicity is classified as being low to high.

Table 10 Resorcinol bis(diphenylphosphate) (RDP, CAS nr 57583-54-7)

	Data	Details	References
Physical–chemical properties			
Molecular weight	574.47 g mol⁻¹		US EPA (2011)
Melting point	90.27ᵃ °C		Hilal et al. (2004)
Water solubility	1.65E−6ᵃ mg L⁻¹	[at 25°C]	Meylan et al. (1996), US EPA (2011)
Water solubility	1.11E−4ᵃ mg L⁻¹	[at 25°C]	US EPA (2011)
Water solubility	6.88E−3ᵃ mg L⁻¹	[at 25°C]	ICL (pers. comm.)
Water solubility	1.05 mg L⁻¹	[at 25°C]	ACD/Labs (2011)
Vapor pressure	5.01E−11ᵃ Pa	[at 25°C]	Hilal et al. (2003)
Vapor pressure	5.29E−7ᵃ Pa	[at 25°C]	US EPA (2011)
Vapor pressure	2.74E−6ᵃ Pa	[at 25°C]	Neely and Blau (1985)
Vapor pressure	2.75E−6ᵃ Pa	[at 25°C]	ICL (pers. comm.)
Vapor pressure	2.59E−3 Pa	[at 20°C]	Meylan and Howard (1991), US EPA (2011)
Henry's law constant	2.98E−8ᵃ Pa m³ mol⁻¹		Hilal et al. (2003)
Henry's law constant	181.7ᵃ Pa m³ mol⁻¹		ICL (pers. comm.)
Log K_{OW}	4.93		ACD/Labs (2011)
Log K_{OW}	5.98ᵃ		Meylan and Howard (1991), US EPA (2011)
Log K_{OW}	7.41ᵃ		Hilal et al. (2004)
Log K_{OW}	11.09ᵃ		US EPA (2011)
Log K_{AW}	−10.92ᵃ		
Persistence			
Low; water	DT_{50} = 7–17 days	20°C, pH 7	European Chemicals Bureau et al. (2007)
Low; water	DT_{50} = 11 days	20°C, pH 4	European Chemicals Bureau et al. (2007), Wildlife International (2000) (not found) from UK Environment Agency et al. (2009a)
Low; water	DT_{50} = 17 days	20°C, pH 7	European Chemicals Bureau et al. (2007), Wildlife International 2000 (not found) from UK Environment Agency et al. (2009a)

(continued)

Table 10 (continued)

	Data	Details	References
Low; water	$DT_{50} = 20$ days	10°C, pH 7, hydrolysis	Wildlife International (2000) (not found) from UK Environment Agency et al. (2009a)
Low; water	$DT_{50} = 21$ days	20°C, pH 9, hydrolysis	Wildlife International (2000) (not found) from UK Environment Agency et al. (2009a)
Low; water	$DT_{50} = 21$ days	20°C, pH 9	European Chemicals Bureau et al. (2007)
Low	Not specified		Clean Production Action et al. (2007)
Low; sludge	Ready biodegradable; $DT_{60} <28$ days		ICL (pers. comm.)
High; water	$DT_{50} = 32$ days	10°C, pH 9, hydrolysis	Wildlife International (2000) (not found) from UK Environment Agency et al. (2009a)
High; water	$DT_{50} = 55$ days	10°C, pH 4, hydrolysis	Wildlife International (2000) (not found) from UK Environment Agency et al. (2009a)
High; water	$DT_{50} = 37.5^a$ days (900 h)	Primary degradation	US EPA (2011)
High; sludge	$DT_{37} = 28$ days		van Ginkel and Stro (1996) (not found) from UK Environment Agency et al. (2009a)
High; soil and sediment	$DT_{50} = 75^a$ days (1,800 h)	Soil, primary degradation	US EPA (2011)
High; soil and sediment	$DT_{50} = 337.5^a$ days (8,100 h)	Sediment, primary degradation	US EPA (2011)
n.c.; atmospheric	$DT_{50} = 12.1\ h^a$	Primary degradation	US EPA (2011)
n.c.; atmospheric	$DT_{50} = 36\ h^e$		Draft UK Environment Agency et al. (2009b)
Bioaccumulation			
Low	Not specified		German Federal Environmental Agency et al. (2001)
Low-high	$BCF = 100–1,000^b$	Based on log $K_{ow} = 4.93$	ICL (pers. comm.)
Moderate	$BCF = 969^{b\ or\ c}\ kg\ L^{-1}$		UK Environment Agency et al. (2009a)
High	Not specified		Clean Production Action et al. (2007)
High	$BCF = 316^b–3,000^{a,b\ or\ c}$		Washington State Department of Ecology and Department of Health (2006), European Chemicals Bureau et al. (2007)

Ecotoxicity

Low	NOEC >1 mg L^{-1}	Daphnia, NOEC exceeds water solubility	ICL (pers. comm.)
Low	LC$_{50}$ = 12.4 mg L^{-1}	Fish	Washington State Department of Ecology and Department of Health (2006), Illinois EPA (2007)
Low	LOEC = 48.64 mg L^{-1}	Algae, growth inhibition	Washington State Department of Ecology and Department of Health (2006)
Low	LC$_{50}$ >100 mg L^{-1}	Daphnia, fish, algae	ICL (pers. comm.)
Low	EC$_{10}$ >121.6 mg L^{-1}	Bacteria	Washington State Department of Ecology and Department of Health (2006)
Moderate	Not specified		Washington State Department of Ecology and Department of Health (2006)
Moderate	Not specified	Chronic exposure (may cause long term effects)	UK Environment Agency et al. (2003), European Chemicals Bureau et al. (2007)
High	EC$_{50}$ = 0.76 mg L^{-1}	Daphnids	Washington State Department of Ecology and Department of Health (2006), Illinois EPA (2007)
High	NOEC = 0.021 mg L^{-1}	Daphnia, 21 days, EC$_{50}$ (immobility) estimated at 0.037 mg L^{-1}, co-solvent used	Wetton and Mullee 2011, unpublished from UK Environment Agency et al. (2009a)

In vivo toxicity

Low	LD$_{50}$ $>2,000$ mg kg^{-1}	Rats, dermal	Washington State Department of Ecology and Department of Health (2006)
Low	LD$_{50}$ $>5,000$ mg kg^{-1}	Rats	Illinois EPA (2007)
Low	LD$_{50}$ $>5,000$ mg kg^{-1} bwt	Rats	Washington State Department of Ecology and Department of Health (2006)
Low	EC$_{50}$ $>20,000$ mg kg^{-1}	Rats	Henrich et al. (2000)
Low	EC$_{50}$ $>1,000$ mg kg^{-1}	Rabbits	Ryan et al. (2000)
Low	Not specified		European Chemicals Bureau et al. (2007)

(continued)

Table 10 (continued)

	Data	Details	References
Moderate	Not specified		UK Environment Agency et al. (2003), Clean Production Action et al. (2007)
Moderate	$LC_{50} = 4.14$ mg L^{-1}	Rats, inhalation	ICL (pers. comm.)
n.c.	*NOEC = 0.1ᵃ mg L⁻¹*	Rats, inhalation (predicts high because of presence TPP, however in vivo toxicity TPP not high)	German Federal Environmental Agency et al. (2001)
In vitro toxicity			
Low	Genotoxicity; mutagenicity	–, *Salmonella* and *E. coli*, AMES test	Washington State Department of Ecology and Department of Health (2006)
n.c.	Endocrine toxicity; metabolization	Metabolites formed: resorcinol diphenylphosphate, hydroxylresorcinol diphenylphosphate, dihydroxyresorcinol diphenylphosphate and hydroxylated parent compounds	Washington State Department of Ecology and Department of Health (2006)

Italic values are predicted: ᵃModeled, ᵇcalculated, ᶜexpert judgment

n.c., not enough data to classify; –, no effects observed

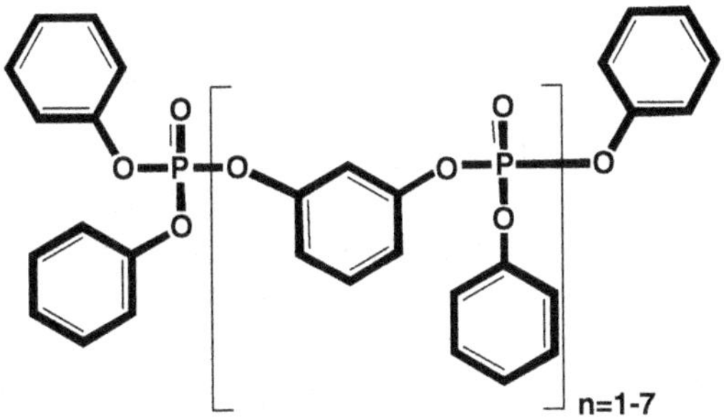

Fig. 2 Schematic
representation of the
chemical structure of the
polymer RDP. The product
typically consists of a
mixture of *oligomers*
with a chain length varying
from 1 to 7

n=1-7

In summary, TPP is a solid with low, but uncertain water solubility and a high potential to partition into lipids. There are many studies available on the PBT properties of TPP. The degradation rate of TPP in air is fast. In water, as well as in sediment and soil, contradictory results on persistence were found. Both long and short dissipation times were reported. Estimations with EPI Suite confirm high persistence in water and soil. The bioaccumulation potential of TPP depends on the species exposed and ranges from low to high values. The ecotoxicity of this HFFR is predominantly moderate to high, whereas other in vivo toxicity is low. For in vitro toxicity, a low genotoxicity and high neurotoxicity were reported. Based on all the observed adverse effects, TPP is labeled as a compound with dangerous effects for the environment by the European Chemicals Agency (ECHA) (General Information on "Labeling" of TPP ECHA Database (Accessed 2011)).

5.2 Resorcinol Bis(diphenylphosphate)

Resorcinol bis(diphenylphosphate) (RDP, CAS 57583-54-7) is a polymeric aryl phosphate and a flame retardant (ICL Industrial Products 2011), typically consisting of a mixture of oligomers having chain lengths between 1 and 7 (Fig. 2). The influence of chain length on the properties of the compound is discussed in the section on APP. Little information is available on the occurrence of RDP in the environment. RDP has been measured in the indoor environment and was present at around 1 ng m^{-3} in domestic air, and at a level of 1,700 ng g^{-1} in house dust (Matsukami et al. 2010). RDP is currently classified as an LPV chemical in the EU (European Chemicals Bureau 2011). For the USA, total annual production was <227 t in 2006 (US EPA 2006).

5.2.1 Physical–Chemical Properties

We assume that all physical–chemical properties reported are based on tests with the technical product (CAS 57583-54-7), because to our knowledge, no purified monomeric RDP is currently available on the market. The monomer of RDP weighs 574.47 g mol^{-1} and for $n=7$, the polymer weighs 2,063.50 g mol^{-1}. RDP is a liquid at room temperature. In Table 10, we show an overview of the physical–chemical properties of this compound. The solubility of RDP in water was measured as being low (1.5 mg L^{-1}, ICL, pers. comm.). Since there are no experimental data for log K_{AW} or log K_{OW}, we estimated values for these properties (Table 10). The vapor pressure, Henry's law constant, and K_{AW} value were all estimated to be low. The log K_{OW} value was predicted to be high, i.e., up to 11.09 (Hilal et al. 2004), showing a preference for partitioning into organic and lipid-rich phases such as soil and biota.

5.2.2 Persistence

Data on RDP degradation in water were contradictory, with examples of low as well as high persistence reported. However, the degradation rate in the atmosphere is fast (see Table 10). One author reported high persistency in sludge (UK Environment Agency et al. 2009a). In agreement, the estimated degradation half-lives using EPI Suite 4.0 (US EPA 2011) indicted a high persistence; i.e., 12.1 h in air, 38 days in water, and more than 75 days in soil or sediment.

5.2.3 Bioaccumulation

There are very few data available on the bioaccumulation of RDP (Table 10). The available references give highly variable data with both high and low bioaccumulation values potential reported. There were little or no details given concerning the test species used or test conditions.

5.2.4 Toxicity

1. Ecotoxicity
   The ecotoxicity of RDP varies between low for some fish, algae, and bacteria (Washington State Department of Ecology and Department of Health 2006; Illinois EPA 2007) and high for daphnids (Washington State Department of Ecology and Department of Health 2006; Illinois EPA 2007) (Table 10).
2. In vivo toxicity
   The toxicity of RDP for rodents is mostly very low, as can be seen in Table 10, with the exception of two studies, in which moderate toxicity was reported
   Two reports mention the presence of TPP as an impurity in RDP (German Federal Environmental et al. 2001; Clean Production Action et al. 2007). The latter quantified the TPP content to be less than 5%. The presence of TPP, as well as the potentially formed toxic TPP-like products, may have an impact on the toxicity of technical RDP, particularly if the exposure is extended.

3. In vitro toxicity

Toxicity data on RDP are scarce. However, as shown in Table 10, no mutagenic effects have been observed. Major fecal metabolites reported are also presented in Table 10. Animal studies do not show adverse biological effects for terato-genic and developmental endpoints at concentrations up to 20,000 mg kg^{-1} diet (Henrich et al. 2000; Ryan et al. 2000). Toxic effects at higher doses could be expected, since commercial RDP contains up to 5% TPP.

In summary, RDP has a low water solubility and high potential for partitioning into organic matter and lipid phases. Considering these characteristics, the high hydrophobicity of this compound and propensity to partition into soil and sediments, RDP might be persistent in the environment; however high as well as low dissipation times were reported. Bioaccumulation is poorly studied and RDP is classified as being low to highly bioaccumulative, although details of the studies cited were not reported. The in vivo (eco)toxicity varies from low to high, with special concern for the TPP impurities present in the commercial product. TPP-like products might also be formed as toxic breakdown products. There is, however, a lack of studies in which this phenomenon has been examined. There are very little data on the in vitro toxicity of RDP, although no mutagenic effects in an AMES test were reported in one study.

5.3 Bisphenol-A Bis(diphenylphosphate)

Bisphenol-A bis(diphenylphosphate) (BDP) is a polymeric aryl phosphate, commonly used as flame retardant (Supresta 2006) and has the CAS registration number 181028-79-5. The technical product (CAS # 5945-33-5) consists of BDP itself (>85%), but its remaining ingredients are largely unknown. According to the Australian Department of Health and Ageing about 0.07% phenol (108-95-2) and <0.01% 4,4'-(1-methylethylidene)bisphenol (80-05-7) are present (Australian Government Regulator of Industrial Chemicals 2000) in the product. These components were not mentioned in another report, in which it was stated that 11% of another phosphoric acid is present (bis[4-[1-[4-[(diphenoxyphosphinyl)oxy]phenyl]-1-methylethyl]phenyl] phenyl ester, CAS #3029-72-5) and <3% of triphenylphosphate (CAS #115-86-6) (Clean Production Action et al. 2007). The bisphenol-A bis(diphenylphosphate) product (CAS number 5945-33-5) will be discussed here to avoid this uncertainty. This is a polymeric compound and typically consists of a mixture of different chain lengths (Fig. 3). As discussed previously, chain length strongly influences the properties of the substance.

Environmental data on BDP are scarce. BDP has been measured in air samples at levels of about 1 ng m^{-3} in domestic indoor sites and at approximately 100 ng g^{-1} in dust (Matsukami et al. 2010). No information is available on production volumes in the EU (European Chemicals Bureau 2011). For the USA, total annual production in 2006 was given as 4,500 to <22,700 t for CAS number 181028-79-5 and an additional 450 to <4,500 t of BDP under the CAS number 5945-33-5 was produced (US EPA 2006).

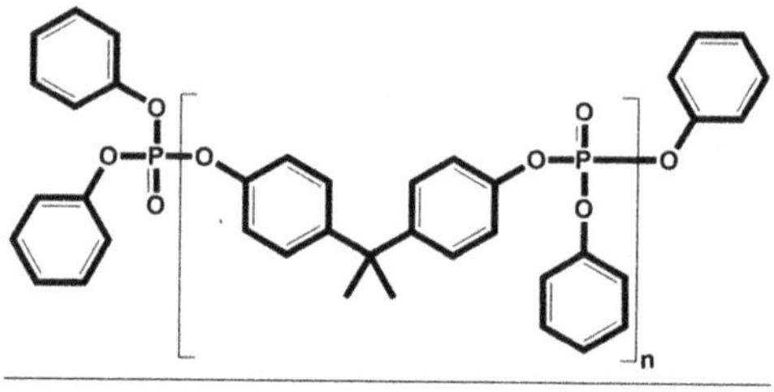

38 S.L. Waaijers et al.

Fig. 3 Schematic representation of the chemical structure of the polymer BDP. The product consists of a mixture of polymers with different chain lengths. The typical composition was unknown

5.3.1 Physical–Chemical Properties

We assume that all physical–chemical properties published for this product are based on tests with the technical product (CAS # 5945-33-5), because to our knowledge no purified monomeric BDP is currently available on the market. The monomer of BDP weighs 693.25 g mol^{-1}, and for $n=10$ it weighs 3,989.80 g mol^{-1}. An overview of its physical–chemical properties is shown in Table 11. BDP is a liquid at room temperature. Mainly low vapor pressure values were reported for the compound (Table 11). However, one author gave a value of 0.18 Torr (approximates to 24 Pa) (Supresta 2006). This latter reference actually reported vapor pressures for a mixture containing >95% BDP and <5% triphenylphosphate. It is therefore expected that this higher reported vapor pressure is less reliable. Additionally, based on the reported low solubility, Henry's law constant and its high molecular weight, it is assumed that the vapor pressure of BDP is also low. Overall, BDP is likely to favor hydrophobic compartments such as soil and biota more than air and water.

5.3.2 Persistence

Data on this chemical are scarce and those available are contradictory, showing high as well as low persistence, with dissipation times ranging from 1 day to 1 year (European Chemicals Bureau et al. 2007). Table 11 provides an overview of the reported data.

5.3.3 Bioaccumulation

There are no experimental studies available on the bioaccumulation of BDP (see Table 11). The few available theoretical studies provided estimates that varied between low and high bioaccumulation. However, it is noteworthy that the study in which a low BCF value was reported, also reported a high log K_{ow} (Table 11). The Australian Department of Health predicted that the BCF for this substance would be high, because of its relatively low molecular weight and water solubility and high log K_{ow} (Australian Government Regulator of Industrial Chemicals 2000). Based on the contradictory persistence data given, we are not fully convinced of the validity of this conclusion. Clearly, there is a need for experimental data to confirm such statements.

Table 11 Bisphenol-A bis(diphenylphosphate) (BDP, CAS nr 5945-33-5)

	Data	Details	References
Physical–chemical properties			
Molecular weight	693.25 g mg L^{-1}		Australian Government Regulator of Industrial Chemicals (2000)
Melting point	41.3–68.6°C		ICL (pers. comm.)
Melting point	40.85–68.85°C		SRC (2006)
Melting point	90°C		US EPA (2011)
Melting point	90.27°C		Hilal et al. (2004)
Water solubility	5.26E−10[a] mg L^{-1}	[at 25°C]	US EPA (2011)
Water solubility	1.92E−7[a] mg L^{-1}	[at 25°C]	US EPA (2011)
Water solubility	1.88E−6[a] mg L^{-1}	[at 25°C]	Eckert and Klamt (2010)
Water solubility	2.27E−4[a] mg L^{-1}	[at 25°C]	SRC (2006)
Water solubility	<1E−3 mg L^{-1}	[at 25°C]	Australian Government Regulator of Industrial Chemicals (2000), ICL (pers. comm.) (latter stated no temperature)
Water solubility	4.15E−1 mg L^{-1}	[at 20°C]	Eckert and Klamt (2010)
Vapor pressure	9.14E−18[a] Pa	[at 25°C]	US EPA (2011)
Vapor pressure	2.74E−6[a] Pa	[at 25°C]	SRC (2006)
Vapor pressure	<1.33E−4 Pa	[at 25°C]	ICL (pers. comm.)
Vapor pressure	1.2E−3 Pa	[at 20°C]	Australian Government Regulator of Industrial Chemicals (2000)
Vapor pressure	<1.2E−3 Pa	[at 25°C]	Hilal et al. (2003)
Vapor pressure	0.12[a] Pa	[at 25°C]	Supresta (2006)
Vapor pressure	24 Pa	[at 25°C]	US EPA (2011)
Henry's Law Constant	4.68E−9[a] Pa m^3 mol^{-1}		SRC (2006)
Henry's Law Constant	5.07E−9 Pa m^3 mol^{-1}		Eckert and Klamt (2010)
Henry's Law Constant	5.38[a] Pa m^3 mol^{-1}		SRC (2006)
Log K_{OW}	4		

(continued)

Table 11 (continued)

	Data	Details	References
Log K_{ow}	≥6		Australian Government Regulator of Industrial Chemicals (2000)
Log K_{ow}	>6		Supresta (2006)
Log K_{ow}	8.79[a]		Eckert and Klamt (2010)
Log K_{ow}	10.02[a]		US EPA (2011)
Log K_{ow}	10[a]		SRC (2006)
Log K_{ow}	14.4[a]		Hilal et al. (2004)
Log K_{AW}	−13.95[a]		Eckert and Klamt (2010)
Log K_{AW}	−11.72[a]		US EPA (2011)
Persistence			
Low	Not specified, primary source not found		Clean Production Action et al. (2007)
Low to high	DT_{50} =1 day to 1 year		European Chemicals Bureau et al. (2007)
High; water	DT_{50} =60 days[a] (1,440 h)		US EPA (2011)
High; water	DT_{50} >1 year	pH 4.0, 7.0, and 9.0. 25°C	European Chemicals Bureau et al. (2007)
High; soil and sediment	DT_{50} =120 days[a] (2,880 h)	Soil	US EPA (2011)
High; soil and sediment	DT_{50} =542 days[a] (13,000 h)	Sediment	US EPA (2011)
High; sludge	Not ready biodegradable (DT_6 =28 days)		Australian Government Regulator of Industrial Chemicals (2000)
n.c.; atmospheric	DT_{50} =0.5 day[a] (12.1 h)		US EPA (2011)
Bioaccumulation			
Low to high?	BCF =3.16[b]	This value is reported with high log K_{ow} so presumably it should be log 3.16 and thus a high value?	Washington State Department of Ecology and Department of Health (2006)
High	Estimated high	high log K_{ow} and relatively low S_w and M_w	Australian Government Regulator of Industrial Chemicals (2000)

Ecotoxicity			
Low to moderate; aquatic	$EC_{50} > 1$ mg L^{-1}	Algae, growth inhibition	Washington State Department of Ecology and Department of Health (2006)
Low to moderate; aquatic	NOEC $= 5$ mg L^{-1}	Fish	Washington State Department of Ecology and Department of Health (2006)
Low to moderate; aquatic	NOEC > 1 mg L^{-1}	Fish, daphnids, and algae (EC_{50} exceeds solubility)	Australian Government Regulator of Industrial Chemicals (2000)
High; aquatic	Not specified	Acute and chronic exposure (bisphenol-A breakdown product is potentially developmentally- and, reproductive toxic, also an endocrine disruptor)	Clean Production Action et al. (2007)
In vivo toxicity			
Low	$LD_{50} > 2,000$ mg kg^{-1} bwt	Rats	Australian Government Regulator of Industrial Chemicals (2000), Washington State Department of Ecology and Department of Health (2006), European Chemicals Bureau et al. (2007)
In vitro toxicity			
Low	Genotoxicity; mutagenicity	–, *Salmonella* and *E. coli*, AMES test	Washington State Department of Ecology and Department of Health (2006)
n.c.	Endocrine toxicity; metabolization	Metabolite: bisphenol-A	Maine Department of Environmental Protection and Maine Center for Disease Control & Prevention (2007)

Italic values are predicted: [a]Modeled, [b]calculated, [c]expert judgment n.c., not enough data to classify; –, no effects observed

5.3.4 Toxicity

1. Ecotoxicity

   The aquatic toxicity of BDP appears to be moderate, although there are only a few poorly described studies available (Table 11). The authors of one study reported the formation of bisphenol-A during testing and the presence of TPP as an impurity. Therefore, the authors of this study concluded that the ecotoxicity of BDP is high, although no experimental details were provided (Clean Production Action et al. 2007).

2. In vivo toxicity

   The in vivo toxicity to rats was described in several studies and was regarded to be low, with a minimum LD_{50} of 2,000 mg kg^{-1} bwt (e.g., Australian Government Regulator of Industrial Chemicals 2000) (Table 11).

3. In vitro toxicity

   An overview of the in vitro toxicity is shown in Table 11. In one study (Maine Department of Environmental Protection and Maine Center for Disease Control & Prevention 2007), it was stated that one of the degradation products of BDP is bisphenol-A, which is an endocrine disrupting compound, but this should be verified as no further details were given. Such incidents make it important to study breakdown products and metabolites.

In summary, BDP has low water solubility. Considering the high hydrophobicity of this compound, it is likely that BDP will accumulate in soil and sediments once released into the environment. Reported persistence ranges from low to high. Bioaccumulation is poorly studied and is estimated to be low to high, although study details were not reported. Ecotoxicity is generally moderate. However high concerns were expressed about the toxicity of the TPP impurity that is present in the commercial product; there was also concern for the potential formation of the toxic breakdown product bisphenol-A (Clean Production Action et al. 2007). However, this phenomenon was not examined in any study. Low in vivo toxicity of BDP was reported in one study. Very limited data exist on the in vitro toxicity of BDP; in the single study available, no mutagenic effects were reported in an AMES test.

5.4 9,10-Dihydro-9-oxa-10-phosphaphenanthrene-10-oxide

The compound 9,10-Dihydro-9-oxa-10-phosphaphenanthrene-10-oxide (DOPO) is currently used as a flame retardant in polymers (MaKuang Chemical Co. Ltd. 2009). There is no information available on production volumes of DOPO (9,10-dihydro-9-oxa-10-phosphaphenanthrene-10-oxide or dihydrooxaphosphaphenantreneoxide, CAS 35948-25-5) from the EU (European Chemicals Bureau 2011) or the USA (US EPA 2006).

5.4.1 Physical–Chemical Properties

DOPO is a solid at room temperature with a melting point between 84.3 and 122°C (Chernysh et al. 1972; Chang et al. 1998; Kuo Ching Chemical Co Ltd. 2009; MaKuang Chemical Co. Ltd. 2009; US EPA 2011). It has a molecular weight of 216.18 g mol^{-1}, and the reported solubility in water varies from moderate and high (0.009–28.97 g L^{-1}, 0.04–134 mol m^{-3}, respectively) (Hilal et al. 2004; Eckert and Klamt 2010; US EPA 2011). An overview of DOPO physical–chemical properties is shown in Table 12. DOPO has a low vapor pressure, Henry's law constant and log K_{AW}, and a moderate log K_{OW} value. It is clear that, once released into the environment, DOPO will not appreciably partition to air. Its properties suggest mostly partitioning to water and a low propensity to partition into soil and biota.

5.4.2 Persistence

In one draft report, it was stated that the chemical is non-persistent (US EPA 2008). The DT$_{50}$ in water is estimated to be shorter than 60 days and in air less than 2 days. However, when using EPI-Suite, the persistence was estimated to be high (US EPA 2011), with a dissipation time (DT$_{50}$) in water, soil, and sediment of more than 37 days each (Table 12). More data are needed to make firm conclusions concerning environmental persistence of this compound.

5.4.3 Bioaccumulation

The estimated BCF for fish is 5.4, and therefore DOPO has been evaluated in this draft reference to be non-bioaccumulative (US EPA 2008).

5.4.4 Toxicity

1. Ecotoxicity
   The aquatic toxicity is estimated to be moderate, since the acute LC$_{50}$ value for algae, considered the most sensitive aquatic species, is estimated at 3 mg L^{-1}, with a reported (draft report) chronic EC$_{50}$ value of 2.4 mg L^{-1} (US EPA 2008). An LC$_{50}$ value for fish of 370 mg L^{-1} was reported by Wetton (1999) in an unpublished draft report (US EPA 2008).
2. In vivo toxicity
   No data are available on the in vivo toxicity of DOPO.
3. In vitro toxicity
   Limited information is available for DOPO. There are no data on carcinogenic, endocrine disruption, or neurotoxicity, although the AMES test showed a negative response for mutagenic activity (Hachiya 1987) (Table 12).

Table 12 Dihydrooxahosphaphenanthrene (DOPO, CAS nr 35948-25-5)

	Data	Details	References
Physical–chemical properties			
Molecular weight	216.18 g mol^{-1}		
Melting point	*84.3^a°C*		US EPA (2011)
Melting point	114–119°C		MaKuang Chemical Co. Ltd. (2009)
Melting point	116–119°C		Kuo Ching Chemical Co Ltd. (2009)
Melting point	117°C		Chernysh et al. (1972)
Melting point	122°C		Chang et al. (1998)
Water solubility	*9.01^a mg L^{-1}*	[at 25°C]	Hilal et al. (2004)
Water solubility	*71.14^a mg L^{-1}*	[at 25°C]	US EPA (2011)
Water solubility	*2,767.1^a mg L^{-1}*	[at 25°C]	US EPA (2011)
Vapor pressure	*8.41E−4^a Pa*	[at 25°C]	Eckert and Klamt (2010)
Vapor pressure	*1.3E−3^a Pa*	[at 25°C]	Hilal et al. (2003)
Vapor pressure	2.93E−3 Pa	[at 25°C]	McEntee (1987)
Vapor pressure	*3.84E−3^a Pa*	[at 25°C]	US EPA (2011)
Henry's law constant	*5.50E−3^a Pa m^3 mol^{-1}*		US EPA (2011)
Henry's law constant	*3.13E−2^a Pa m^3 mol^{-1}*		Hilal et al. (2003)
Log K_{OW}	*1.18^a*		Eckert and Klamt (2010)
Log K_{OW}	*1.87^a*		US EPA (2011)
Log K_{OW}	*3.32^a*		Hilal et al. (2004)
Log K_{AW}	*−5.65^a*		US EPA (2011)
Log K_{AW}	*−8.6^a*		Eckert and Klamt (2010)
Persistence			
Low; water	*$DT_{50} < 60^{a,e}$ days*		US EPA (2008)
High; water	*$DT_{50} = 37.5^a$ days*	Primary degradation, modeled	US EPA (2011)
High; soil and sediment	*$DT_{50} = 75^a$ days*	Soil, primary degradation, modeled	US EPA (2011)
High; soil and sediment	*$DT_{50} = 337.5^a$ days*	Sediment, primary degradation, modeled	US EPA (2011)
n.c.; atmospheric	*$DT_{50} < 2^{a,e}$ days*		US EPA (2008)
n.c.; atmospheric	*$DT_{50} = 43^a$ h*	Primary degradation, modeled	US EPA (2011)
Bioaccumulation			
Low	*$BCF = 5.4^{a,e}$*	Fish	US EPA (2008)
Ecotoxicity			
Low	$LC_{50} = 370^e$ mg L^{-1}	Fish, 48 h	Wetton (1999) unpublished from US EPA (2008)
Low	*$LC_{50} = 230^{a,e}$ mg L^{-1}*	Daphnids, 48 h	US EPA (2008)
Low	*$EC_{50} = 23^{a,e}$ mg L^{-1}*	Daphnids, chronic	US EPA (2008)
Low	*$LC_{50} = 20^{a,e}$ mg L^{-1}*	Fish, 96 h	US EPA (2008)
Low	*$EC_{50} = 16^{a,e}$ mg L^{-1}*	Fish, chronic	US EPA (2008)
Moderate	*$EC_{50} = 3^{a,e}$ mg L^{-1}*	Algae, 96 h	US EPA (2008)
Moderate	*$EC_{50} = 2.4^{a,e}$ mg L^{-1}*	Algae, chronic	US EPA (2008)
In vitro toxicity			
Low	Genotoxicity; mutagenicity	–	Hachiya (1987)

Italic values are predicted: [a]Modeled, [b]calculated, [c]expert judgment

n.c., not enough data to classify; –, no effects observed

[e]Not all primary sources are found from (US EPA 2008), also this reference is a draft report, so reported values may be not final

In summary, DOPO is a solid at room temperature with a moderate to high water solubility. There is a lack of experimental persistence data. Modeled values of the persistence in water, which is probably the most relevant environmental compartment for DOPO, are contradictory and range from high to low; obviously more research is needed. Bioaccumulation of DOPO was estimated to be low in one study. Ecotoxicity was reported to range from moderate to low in the same study. There are not enough data to classify DOPO's in vivo toxicity. Only one study showed a negative response for mutagenic activity.

5.5 Aluminum Diethylphosphinate

Aluminum diethylphosphinate (ALPI, CAS # 225789-38-8) is a metal phosphinate salt and used as a flame retardant in epoxies and polymers (Clariant 2007). No information is available on production volumes in the EU (European Chemicals Bureau 2011) or the USA (US EPA 2006).

5.5.1 Physical–Chemical Properties

ALPI is solid at room temperature. A melting point has not been reported, probably because ALPI is reported to decompose at temperatures ranging from 400°C (Australian Government Regulator of Industrial Chemicals 2005). It has a reported solubility in water of 2.5 g L^{-1} (2.56E–3 to 6.41 mol m^{-3}) (Clariant 2007). ALPI has a low vapor pressure and K_{OW}, presumably favoring the water phase over other compartments once released into the environment. An overview of its properties is shown in Table 13.

5.5.2 Persistence

As mentioned previously in the introduction, persistence expressed as dissipation times is not considered to be very relevant for metals, in this case aluminum. However, the counter ion diethylphosphinate is organic. Three references were found in which a moderate to high persistence of ALPI was claimed (Table 13).

5.5.3 Bioaccumulation

In a draft report, the US EPA predicted that the BCF value for ALPI is <1,000, meaning that it has a low bioaccumulation potential (US EPA 2008). A low bioaccumulation potential was reported in three other studies, although no detailed information or specific data were provided (Danish EPA et al. 2007; European Chemicals Bureau et al. 2007; Dekant 2009) (Table 13).

Table 13 Aluminum diethylphosphinate (ALPI, CAS nr 225789-38-8)

	Data	Details	References
Physical–chemical properties			
Molecular Weight	390.3 g mol^{-1}		
Melting point	>400°C (decomposition)		Australian Government Regulator of Industrial Chemicals (2005)
Water solubility	2,500 mg L^{-1}	[at 25°C]	Clariant (2007)
Log K_{OW}	−0.44		Clariant (2007)
Persistence			
Moderate	*Days to weeks*[a, c, e]		US EPA (2008)
High; sludge	Not inherently biodegradable	OECD Guideline 302C (Inherent Biodegradability: Modified MITI Test (II)), aerobic, 28 days	European Chemicals Agency (ECHA) Database (original study 2009b)
High; sludge	Not ready biodegradable	OECD Guideline 301F (Ready Biodegradability: Manometric Respirometry Test), aerobic 28 days	Clariant (2007), Danish EPA et al. (2007), European Chemicals Bureau et al. (2007), European Chemicals Agency (ECHA) Database (original study 2009c)[d]
Bioaccumulation			
Low	*BCF <1,000*[,g]		US EPA (2008)
Low	Not specified		Danish EPA et al. (2007), European Chemicals Bureau et al. (2007), Dekant (2009)
Ecotoxicity			
Low	LC_{50} >9.2[e] mg L^{-1} LC_{50} >11[e] mg L^{-1}	Zebra fish, 96 h	US EPA (2008)
Low	EC_{50} =60–76 mg L^{-1}	Algae, *Scenedesmus subspicatus*, 72 h, pH not adjusted	Clariant (2007)[d]
Low	EC_{50} =50–76 mg L^{-1e}	Algae, 72 h	US EPA (2008)
Low	LC_{50} >33 mg L^{-1e}	Daphnids, 48 h	US EPA (2008)
Low	EC_{50} =ca. 46.2 mg L^{-1}	*Daphnia magna*, chronic, reproduction, LOEC=ca.32 mg L^{-1} NOEC=10 mg L^{-1}	US EPA (2008), ECHA Database original study (2005-Apr-13)
Low	EC_{50} =48[a,e] mg L^{-1}	Fish, chronic	US EPA (2008)
Low	EC_{50} >100 mg L^{-1}	*Daphnia magna*, 48 h, EC_{50} exceeds solubility	ECHA Database original study (1998a)

Rating	Value	Description	Reference
Low	$EC_{50} > 100$ mg L^{-1}	Zebrafish, EC_{50} exceeds solubility	ECHA Database original study (1998b)
Low	$EC_{50} > 100$ mg L^{-1}	Zebrafish, chronic, 28 days (nominal concentration)	ECHA Database original study (2009a)
Low	NOEC > 180 mg L^{-1}	Algae	European Chemicals Agency (ECHA) Database original study (1998c)[d]
Low	$EC_{50} = 1,968$ mg L^{-1}	OECD Guideline 209 (Activated Sludge, Respiration Inhibition Test), 3 h, NOEC = 483 mg L^{-1}, nominal concentrations	ECHA Database original study (1998d)
Moderate	ChV = 1.8 mg L^{-1}	Algae, chronic exposure, ChV = chronic value, no further details provided	Draft US EPA (2008)
Moderate	*ChV = 1.4*[a, b or c, s] *mg L^{-1}*	Algae, chronic exposure, ChV = chronic value, no further details provided	Draft US EPA (2008)
In vivo toxicity			
Low	NOEC > 1 g kg^{-1} bwt day^{-1}	Rats, repeated dose study, 28 days	ECHA Database original study (1998g)
Low	NOEC = 1 g kg^{-1} bwt day^{-1}	Rats, reproduction toxicity, minor changes in both sexes at the highest dosage of 1 g kg^{-1} bwt day^{-1}, ~48 days	ECHA Database original study (2008a, b)
Low	$LD_{50} > 2$ g kg^{-1} bwt	Acute oral toxicity	US EPA (2008), ECHA Database original study (1998b)[d], first is draft[e]
Low	$LD_{50} > 2$ g kg^{-1} bwt	Acute dermal toxicity	ECHA Database original study (1998a)
Low	Non-mutagenic in vivo	No mutagenic effect up to 2 g kg^{-1} bwt, OECD Guideline 474 (Mammalian Erythrocyte Micronucleus Test)	ECHA Database original study (2008a, b)
In vitro toxicity			
Low	Genotoxicity; mutagenicity	–, Salmonella, AMES test	European Chemicals Bureau et al. (2007)
Low	Genotoxicity; mutagenicity	–, OECD 471 (Bacterial Reverse Mutation Assay)	ECHA Database original study (1998e)[d]
Low	Genotoxicity; mutagenicity	–, OECD Guideline 473 (In vitro Mammalian Chromosome Aberration Test), tested up to 0.780 mg mL^{-1}	ECHA Database original study (1998f)

Italic values are predicted: [a]Modeled, [b]calculated, [c]expert judgment

n.c., not enough data to classify; –, no effects observed

[d]A recent publication (Danish EPA et al. 2007) also discusses the PBT data of Alpi, this is predominantly based on the same reports from Clariant (producer) as referred to in this review

[e]Not all primary sources are found from (US EPA 2008), also this reference is a draft report, so reported values may be not final

5.5.4 Toxicity

When ALPI is dissolved in water, the complexation of the ions will depend on the prevailing conditions. A low pH, for example, Al^{3+} dissolution will be favored. Complexation and speciation are not treated in detail in this review. Freely dissolved aluminum ($Al^{3+}(aq)$) can be highly toxic. The toxicity of aluminum has been studied extensively (Berthon 2002) (Kucera et al. 2008) and is not addressed here.

1. Ecotoxicity
   A few studies are available on the aquatic toxicity of ALPI, on algae, daphnids, crustaceans, and fish, and these generally show a low to moderate toxicity (Table 13).
2. In vivo toxicity
   Several low NOEC and LC_{50} values that were >1 $g\,kg^{-1}\,day^{-1}$ were reported for in vivo rodent toxicity for ALPI in two studies (Table 13).
3. In vitro toxicity
   Although toxicity data on aluminum and several aluminum compounds are available, data on ALPI were limited (for a review see (The Subcommittee on Flame-Retardant Chemicals 2000)). No mutagenic activity was observed in the AMES test, with or without metabolic activation (European Chemicals Bureau et al. 2007). In vitro toxicity of ALPI is classified as low.

In summary, ALPI is a solid at room temperature with moderate water solubility. Persistence was reported as being moderate to high; however, this was based on three studies in which few details were provided. The bioaccumulation of ALPI was estimated to be low, although there was also a scarcity of studies. The ecotoxicity of ALPI is low, with the exception of one study in which moderate toxicity for algea was reported. For in vivo toxicity, two studies are available and indicate low toxicity to rodents. The in vitro mutagenicity of ALPI was classified as low; no other in vitro toxicity data were available.

6   A Nitrogen-Based Organic Flame Retardant: Melamine Polyphosphate

Melamine polyphosphate (MPP, CAS 218768-84-4) is a nitrogen-based organic salt and is used as a flame retardant (Budenheim 2010; Ciba 2010). The compound dissociates into melamine and polyphosphoric acid in water. There is no information on the environmental presence available for MPP, nor is there information available on production volumes in the EU (European Chemicals Bureau 2011) or the USA (US EPA 2006).

6.1 Physical–Chemical Properties

As is the case for APP, MPP is chemically synthesized as an ionic polymer. It is thus a mixture of polymers having different chain lengths or degrees of branching, and accordingly the physical–chemical properties change with those factors as well (as discussed in section *Ammonium polyphosphate (APP)*). MPP was reported to decompose at 400°C or higher (Australian Government Regulator of Industrial Chemicals 2006). The solubility in water was reported to be high, i.e., <100 g L^{-1}, i.e., miscible (Nordin 2007; Eckert and Klamt 2010). Its physical–chemical properties are shown in Table 14.

Table 14 Melamine polyphosphate (MPP, CAS nr 218768-84-4)

	Data	Details	References
Physical–chemical properties			
Molecular weight	>10,000 g mol^{-1}		Australian Government Regulator of Industrial Chemicals (2006)
Melting point	>400°C		Australian Government Regulator of Industrial Chemicals (2006)
Water solubility	<100 mg L^{-1}	[at 22°C]	Nordin (2007)
Water solubility	*Miscible*[a]		Eckert and Klamt (2010)
Vapor pressure	6.65E–3 Pa	[at 25°C]	Eckert and Klamt (2010)
Vapor pressure	<<8 Pa	[at 25°C]	Australian Government Regulator of Industrial Chemicals (2006)
Vapor pressure	<8 Pa	[at 25°C]	Nordin (2007)
Log K_{OW}	<−2.3	[at 20°C]	Australian Government Regulator of Industrial Chemicals (2006)
Log K_{OW}	−2.3	[at 25°C]	Nordin (2007)
Log K_{OW}	*−2.15*[a]	[at 25°C]	Eckert and Klamt (2010)
Log K_{AW}	*−10.9*[a]	[at 25°C]	Eckert and Klamt (2010)
Persistence			
High	DT$_{50}$ = weeks to months[e] for melamine		US EPA (2008)
Bioaccumulation			
Low	BCF < 3.8[e]	*Cyprinus carpio*, melamine	US EPA (2008)
Low	*BCF < 1,000*[c,e]		US EPA (2008)
Low	Not specified		Australian Government Regulator of Industrial Chemicals (2006)

(continued)

Table 14 (continued)

	Data	Details	References
Ecotoxicity			
Low	$EC_{50} > 3.0^e$ mg L^{-1}	Algae, melamine	US EPA (2008)
Low	$EC_{50} > 3.0$ mg L^{-1}	Algae	Australian Government Regulator of Industrial Chemicals (2006), European Chemicals Bureau et al. (2007)
Low	$EC_{50} = 940^e$ mg L^{-1} NOEC $= 320.0^e$ mg L^{-1}	Algae, melamine, 96 h	US EPA (2008)
Low	$EC_{50} = 32$–56^e mg L^{-1}	Daphnids, melamine, chronic exposure	US EPA (2008)
Low	$EC_{50} > 2,000^e$ mg L^{-1}	Dapnnids, melamine, 48 h	US EPA (2008)
Low	$LC_{50} > 500^e$ mg L^{-1}	Fish, melamine	US EPA (2008)
Low	$EC_{50} = 1,000$–$3,000^e$ mg L^{-1}	Fish, melamine, chronic	US EPA (2008)
In vivo toxicity			
Low	$LD_{50} = 4,000^e$ mg kg^{-1} bwt	Polyphosphoric acid, rats	US EPA (2008)
Low	$LD_{50} = 3,160$–$7,014^e$ mg kg^{-1} bwt	Melamine: rats, mice	US EPA (2008)
Low	$LD_{50} > 1,000^e$ mg L^{-1}	Rabbits, dermal	US EPA (2008)
Low	$LD_{50} = 3,248^e$ mg L^{-1}	Rats, inhalation	US EPA (2008)
Low	$LD_{50} > 2,000$ mg kg^{-1} bwt	Rats, melamine polyphosphate (incl. different technical products tested)	Australian Government Regulator of Industrial Chemicals (2006), Nordin (2007), Budenheim (2010), Ciba (2010)
In vitro toxicity			
n.c.	Genotoxicity; carcinogenicity	$\pm^e$	US EPA (2008)

Italic values are predicted: [a]Modeled, [b]calculated, [c]expert judgment
n.c., not enough data to classify; ±, toxicity effects observed
[d]The report refers to an OECD SIDS
[e]Not all primary sources are found from (US EPA 2008), also this reference is a draft report, so reported values may be not final

6.2 Persistence

In one draft report, it was stated that MPP has a high persistence, namely DT_{50} >1 year for polyphosphoric acid, and a DT_{50} of weeks to months for melamine (US EPA 2008).

6.3 Bioaccumulation

In a draft report, the EPA predicted that the BCF of MPP to be less than 1,000, meaning that it has a low bioaccumulation potential (US EPA 2008). Another author reported a low bioaccumulation potential, but gave no data (Australian Government Regulator of Industrial Chemicals 2006) (Table 14).

6.4 Toxicity

1. Ecotoxicity
   In the few studies available, low ecotoxicity of MPP to algae, daphnids, and fish were reported, with EC_{50} values of 3.0 mg L^{-1} or higher (algae) (Australian Government Regulator of Industrial Chemicals 2006; European Chemicals Bureau et al. 2007) as can be seen in Table 14.
2. In vivo toxicity
   The in vivo toxicity of MPP appears to be low as well, only showing effects to rodents when they are exposed to the compound at >1,000, and up to 4,000 mg kg^{-1} bwt (Table 14).
3. In vitro toxicity
   Information on in vitro toxicity of MPP is limited but, based on the toxicity of melamine, it is expected that MPP has low hazard for carcinogenicity (Illinois EPA 2007). In contrast, studies with MPP showed effects for in vivo chromosomal aberration tests and in vivo sister chromatic assays with mice in a draft report (US EPA 2008). There are not enough data to classify the in vitro toxicity of MPP.

In summary, MPP is a solid at room temperature with high solubility in water. Once dissolved in water, it will dissociate into melamine and polyphosphoric acid.

A high persistence of phosphoric acid was reported once. The bioaccumulation potential of MPP is low, although no details were provided. In vivo (eco)toxicity is low. There are not enough data to classify the in vitro toxicity.

7 An Intumescent System: Pentaerythritol

Almost all intumescent systems consist of three basic components, a dehydrating component, such as ammonium polyphosphate, a charring component, such as pentaerythritol (PER) and a gas source or blowing agent, often a chemical containing nitrogen such as melamine polyphosphate or ammonium polyphosphate (ENFIRO 2009). The latter two are reviewed in the sections *Melamine polyphosphate* and *Ammonium polyphosphate*, respectively.

Production volumes of pentaerythritol (CAS 115-77-5) in Japan were 24,074 t in 1996 and 27,513 t in 1997 (UNEP OECD SIDS 1998, 2005). PER is classified as an HPV chemical in the EU (European Chemicals Bureau 2011). Total annual production was given for the USA as 45,000 to <227,000 t in 2006 (US EPA 2006).

7.1 Physical–Chemical Properties

Pentaerythritol is a white crystalline solid at environmental temperatures and its melting point is 260°C (Hilal et al. 2003; Syracuse Research Corporation (SRC) 2009). It has a high solubility in water (Table 15). PER is a highly water soluble compound with a relatively low vapor pressure, low Henry's Law constant, low $\log K_{OW}$, and low $\log K_{AW}$. Its properties are reported in Table 15.

7.2 Persistence

PER has shown a high as well as low persistence in different degradation experiments (Table 15). Since PER is hydrophilic, the dissipation times in water (and sludge) seem most relevant. In a few studies, it was stated that this compound degrades very slowly in sludge and water. However, more studies show that PER degrades quickly, for example, a ring test from 1985 showed a DT_{60} of 28 days (European Chemicals Agency (ECHA) Database original study 1985).

7.3 Bioaccumulation

PER has a low bioaccumulation factor of 0.3–0.6 for the fish species *Cyprinus carpio* (European Chemicals Bureau 2000e). Since it has a very low $\log K_{OW}$ value, its bioaccumulation potential is also likely to be low (Table 15).

7.4 Toxicity

1. Ecotoxicity
   The ecotoxicity of PER has been studied with several species, i.e., fish, algae, daphnia, and bacteria. PER has low toxicity to all species, with EC_{50} values over 100 mg L^{-1} (European Chemicals Bureau 2000e) (Table 15).
2. In vivo toxicity
   A low toxicity of PER to several rodent species was reported in two studies, with an LD_{50} >2,000 mg kg^{-1} bwt (ECHA Database original study 1996-Jul-25) (Table 15).

Table 15 Pentaerythritol (PER, CAS nr 115-77-5)

	Data	Details	References
Physical–chemical properties			
Molecular weight	136.15 g mol^{-1}		US EPA (2011)
Melting point	74°C		US EPA (2011)
Melting point	119°C		ECHA Database original study (2009e)
Melting point	258°C		US EPA (2011)
Melting point	258°C		Hilal et al. (2003), SRC (2009)
Melting point	260°C		Perstorp Specialty Chemicals AB (2008), ECHA Database original study (2009f)
Water solubility	62,000 mg L^{-1}	[at 20°C]	Meylan and Howard (1995)
Water solubility	72,300 mg L^{-1}	[at 25°C]	Hilal et al. (2004)
Water solubility	3.93e+5[a] mg L^{-1}	[at 25°C]	US EPA (2011)
Water solubility	1e+6[a] mg L^{-1}	[at 25°C]	Eckert and Klamt (2010)
Water solubility	3.65e+6[a] mg L^{-1}	[at 25°C]	Neely and Blau (1985)
Vapor pressure	3.37E-6 Pa		US EPA (2011)
Vapor pressure	3.38E-6 Pa		ECHA Database original study (2010-Jan-31)
Vapor pressure	3.4E-6[a] Pa		ECHA Database original study (2010-Jan-31)
Vapor pressure	15E-6[a] Pa		Hilal et al. (2003)
Vapor pressure	1.62E-5[a] Pa		Perstorp Specialty Chemicals AB (2008)
Vapor pressure	<1E-3 Pa		Eckert and Klamt (2010)
Vapor pressure	2.95E-2[a] Pa		Hilal et al. (2003)
Henry's Law constant	1.13E-9[a] Pa m^3 mol^{-1}		US EPA (2011)
Henry's Law constant	4.15E-5[a] Pa m^3 mol^{-1}		(Meylan and Howard 1991)
Log K$_{OW}$	-4.15[a]		Hilal et al. (2004)
Log K$_{OW}$	-1.77[a]		US EPA (2011)
Log K$_{OW}$	-1.70		Perstorp Specialty Chemicals AB (2008)
Log K$_{OW}$	-1.70[b]		ECHA Database original study (2009d)
Log K$_{OW}$	-1.69		SRC (2011)

(continued)

Table 15 (continued)

	Data	Details	References
Log K_{OW}	-1.4^a		Eckert and Klamt (2010)
Log K_{AW}	-9.35^a		Eckert and Klamt (2010)
Log K_{AW}	-7.78^a		US EPA (2011)
Persistence			
Low; water	$DT_{50} = 208^a$ h (8.6 days)	Primary degradation, modeled	US EPA (2011)
Low; water	$DT_{60} < 28$ days	Ring test, 25 studies	ECHA Database original study (1985)
Low; water	Biodegradable	Biodegradable under some test conditions, no details provided	ECHA Database original study (1980-Feb-29 & 1981)
Low; sludge	$DT_{84} = 28$ days (3.87 S.D.), ready biodegradable		ECHA Database original study (1991-Sep-16)
Low; sludge	Ready biodegradable	99% DOC removal in 28 days	ECHA Database original study (1990-Sep-24)
Low; sudge	Inherent biodegradable	99% DOC removal in 28 days	ECHA Database original study (1994-Jan-25)
Low; sludge	Inherent biodegradable	>90% DOC removal, few details provided	ECHA Database original study (1979-Feb-08)
High; water	No hydrolysis	At 20°C after 5 days for pH 5, 7, 9	ECHA Database original study (2010-Aug-2)
High; water	$DT_{13} = 25$ days		European Chemicals Bureau (2000e)
High; soil and sediment	$DT_{50} = 416^a$ h	Soil, primary degradation, modeled	US EPA (2011)
High; soil and sediment	$DT_{50} = 1,870^a$ h	Sediment, primary degradation, modeled	US EPA (2011)
High; sludge	$DT_{50} = 14$ to more than 28 days	Aerobic sewage sludge	European Chemicals Bureau (2000e)
High; sludge	Not ready biodegradable		European Chemicals Bureau (1998 (2005 updated))
n.c.; atmospheric	$DT_{50} = 15.4^a$ h	Primary degradation, modeled	US EPA (2011)
Bioaccumulation			
Low	BCF = 0.3–0.6	Fish, *Cyprinus carpio*	European Chemicals Bureau (2000e)
Ecotoxicity			
Low	$LC_{50} > 100$ mg L^{-1}	(Nominal concentration) *Oryzias latipes*, fresh water, 96 h	ECHA Database original study (1993-Jan-03)
Low	$EC_{50} > 1,000$ mg L^{-1}	(Nominal concentration) *Daphnia magna*, 24 h	ECHA Database original study (1993-March-01a)

Low	$NOEC = 1,000$ mg L^{-1}	(Nominal concentration) *Daphnia magna*, 21 days	ECHA Database original study (1993-March-01b)
Low	$NOEC = 1,000$ mg L^{-1}	(Nominal concentration) Microorganisms (activated sludge from domestic sewage)	ECHA Database original study (2010-April-29)
Low	$EC_{10} > 500$ mg L^{-1}	*Pseudomonas putida* (nominal concentration)	Knie et al. (1983) (not found) from ECHA Database Original Study (1983)
Low	$LD_{50} > 5,000$ mg L^{-1}	Fresh water fish	European Chemicals Bureau (2000e)
Low	$EC_{50} > 500$ to $>>5,000$ mg L^{-1}	Crustaceans, algae, and bacteria	European Chemicals Bureau (2000e)
Low	$IC_{50} = 600$ mg L^{-1}	Daphnids	European Chemicals Bureau (1998 (2005 updated))

In vivo toxicity

Low	$NOEC = 100$ mg kg bwt^{-1} day^{-1}	Rats, repeated dose study, ~46 days	ECHA Database original study (1996-Jul-28)
Low	$NOEC = 1,000$ mg kg bwt^{-1} day^{-1}	Rats, repeated dose study, 28 days	ECHA Database original study (1992-May-30)
Low	$NOEC = 1,000$ mg kg bwt^{-1} day^{-1}	Rats, repeated dose study, ~46 days, developmental and reproductive toxicity	ECHA Database original study (1996-Jul-28)
Low	$LD_{50} = low$	Rats survived doses as high as 16,000 mg kg^{-1} bwt, few details	ECHA Database original study (1964)
Low	$LD_{50} > 2,000$ mg kg^{-1} bwt	Rats	ECHA Database original study (1996-Jul-25)
Low	$LD_{50} > 5,110$ mg kg^{-1} bwt	Rats	ECHA Database original study (1990)
Low	$LC_{50} > 10,000$ mg kg^{-1}	Rabbits	ECHA (1963-May-19)
Low	$LC_0 = 11,000$ mg m^{-3} air	Rats, inhalation, 6 h exposure (LC_0, no effect)	ECHA Database original study (1964a, b)
Low	$LD_{50} = 11,300$ mg kg^{-1} bwt	Guinea pigs	ECHA Database original study (1964a, b)
Low	$LD_{50} = 25,500$ mg kg^{-1} bwt	Mice	ECHA Database original study (1964a, b)
Low	$LD_{50} > 10,000$ mg kg^{-1} bwt	Rats, mice, rabbits, and guinea pigs	Merck Chemicals—Product Information (Merck Website), European Chemicals Bureau (2000e)

(continued)

Table 15 (continued)

Data	Details	References
In vitro toxicity		
Low Genotoxicity; mutagenicity	–	UNEP OECD SIDS (1998, 2005)
Low Genotoxicity; mutagenicity	–, Ames test negative, tested up to 5,000 µg plate^{-1}, OECD Guideline 471 (Bacterial Reverse Mutation Assay)	Shimizu et al. (1985), ECHA Database original study (1994-Dec-07)
Low Genotoxicity; mutagenicity	–, OECD Guideline 473 (In vitro Mammalian Chromosome Aberration Test)	ECHA Database original study (1994-Dec-21)
Low Genotoxicity; mutagenicity	–, OECD Guideline 476 (In vitro Mammalian Cell Gene Mutation Test)	ECHA Database original study (2010-Okt-2)
Low Genotoxicity; mutagenicity	–, OECD Guideline 471 (Bacterial Reverse Mutation Assay), tested up to 5,000 µg plate^{-1}	Shimizu et al. (1985)

Italic values are predicted: [a]Modeled, [b]calculated, [c]expert judgment

n.c., not enough data to classify; –, no effects observed

3. In vitro toxicity

Information on the in vitro toxicity of PER is limited (Table 15). However, in several studies no mutagenicity was observed. One author reported that reproduction and developmental studies did not show any toxicity up to 1,000 mg kg^{-1} day^{-1} (UNEP OECD SIDS 1998, 2005).

In summary, PER is a solid at room temperature and is highly water soluble. In most studies, a low persistence was reported, although examples of high values were also reported. A low BCF value was reported for fish. The in vivo (eco)toxicity of PER appears to be low. There are not enough data to classify the in vitro toxicity of PER.

8 Discussion

8.1 Data Availability

Ideally, the HFFRs that replace the existing halogenated flame retardants should pose lower risks to the environment and to human health. Yet, our review revealed that, apart from TPP, all potentially replacement compounds have large data gaps concerning their published PBT properties. Indeed, for some of these compounds, even the most basic physical–chemical properties have not yet been disclosed. Because these compounds are currently produced and distributed on a global scale, in some cases even as HPV chemicals, it is crucial to fill these data gaps. It is conceivable that, with the implementation of REACH, more data may become or are already available on these compounds. Dossiers with information on PBT properties may exist, for instance in the US EPA and the ECHA archives. If so, then these were not publically accessible. Data for some compounds have recently become available on the ECHA website (Registered substances http://echa.europa.eu/web/guest/information-on-chemicals/registered-substances), but we only noted this after we completed our literature review at the end of August 2011. The availability of such information could substantially contribute to filling the presently identified data gaps and would greatly accelerate the risk evaluation of the compounds addressed in this review; these data are also needed because these compounds are currently being marketed.

Despite the REACH regulations, characterization of compounds often lacks important in-depth studies, such as the identification and characterization of potentially toxic metabolites or decomposition products. This is one of the reasons that the European Commission has funded a research project on HFFRs, called ENFIRO (A Life Cycle Assessment of Environment-Compatible Flame Retardants (Prototypical Case Study)). ENFIRO aims to fill a large part of the existing data gaps identified in the present review. ENFIRO is studying several aspects, including environmental and toxicological risks, fire safety, product application, and viability of industrial implementation (ENFIRO 2009). If successful, a solid basis will be formed for assessing the suitability of the HFFRs as safe and environmentally friendly alternatives.

8.2   Inconsistency of Data

We demonstrated in the present review that for many of the HFFRs, widely different values for the same properties have been published in literature. When assessing data quality, we considered the experimental values to be more reliable than the modeled ones; modeled values were, in turn, considered more reliable than the so-called expert judgments. Consequently, we preferred data published in peer-reviewed scientific papers over those in reports and other so-called grey literature. Perhaps the most important issue was the transparency of the experimental setup; the more detail that researchers provide on the test conditions and results, the better. Differences in prevailing conditions and methods may explain the observed differences between test results. For example, a low pH may favor degradation by hydrolysis or, in the case of a metal salt, the toxicity may change dramatically (Peterson et al. 1984; Martinez and Motto 2000; Spurgeon et al. 2006). Additionally, the purity and composition of the products tested is often not reported. Possibly, the technical products used for the experiments vary in polymer formulations, e.g., coated vs. uncoated forms, leading to different results in reported PBT properties.

8.3   Persistence, Bioaccumulation, and Toxicity of the Selected HFFR

An overview of the classification of the selected compounds that is based on the REACH criteria for PBT and vPvB chemicals is given in Table 16. It is important to realize though, that these assessments are truncated, and data presented in the relevant sections should be consulted for the detailed data. In particular, bioaccumulation and toxicity are species dependent, and even variations among individuals within the same species are not uncommon (Baird et al. 1989). Therefore, it is not surprising that high as well as low classifications sometimes were reported for the same parameter. Furthermore, bioaccumulation was a more difficult parameter to assess, because many studies did not consider depuration times of the chemical. In some toxicity experiments, carrier solvents were used for poorly water soluble organic compounds. Such solvents may enhance exposure concentrations of the tested compounds that exceed their water solubility, which then undermines environmental relevance of the data. Water solubility is often difficult to assess anyway, because reported water solubility values can range over several orders of magnitude. Nevertheless, we based our assessments on the reported effect concentrations.

Three HFFRs immediately drew our attention (Table 16): TPP, RDP, and BDP. TPP has been studied quite extensively and is clearly persistent, bioaccumulative, and toxic. Because they have high to low reported ecotoxicity and persistence, neither RDP nor BDP seem to have proven to be promising alternative flame retardants yet, but this view is based on a limited number of studies. Details on the bioaccumulation potential of RDP and BDP were not provided, although both were classified

Table 16 Overview of PBT properties for selected HFFRs

Compound	Persistence	Bioaccumulation	vPvB?	Toxicity Ecotoxicity	In Vivo toxicity	In vitro toxicity	Overview page nr
TPP	Low to high	Low to high	(Yes)	Low to high	Low	Low to high	Table 9
RDP	Low to high	Low to high	No	Low to high	Low (to moderate)	(Low)	Table 10
BDP	Low to high	(Low to high)	(No)	Low to high	(Low)	(Low)	Table 11
ATH	–	(Low)	(no)	Low to high	(Low)	(Low)	Table 3
ZB	–	n.d.	n.d.	High	Low to high	(Low)	Table 6
ALPI	Moderate to high	(Low, not specified)	(No)	Low to moderate	Low	Low	Table 13
PER	Low to high	(Low)	(No)	Low	Low	Low	Table 15
DOPO	(Low to high)	(Low)	(No)	Low to moderate	n.d.	(Low)	Table 12
MPP	(High)	(Low)	(No)	Low	Low	n.d.	Table 14
APP	–	(Low, not specified)	(No)	Low (to moderate)	(Low)	(Low)	Table 5
ZHS	–	(Low, not specified)	(No)	Low	Low (to moderate)	(Low)	Table 7
Mg(OH)$_2$	–	n.d.	n.d.	n.d.	(Low)	n.d.	Table 4
ZS	–	(Low, not specified)	(low)	(low)	(Low)	n.d.	Table 8

Please note that this table gives an overview of the data found in literature and it is not an assessment. For a more detailed clarification refer to the corresponding paragraphs

(Bracketed) = based on two or less studies, n.d. = no data

as potentially highly bioaccumulative. Clearly, there is a need for research to clarify these uncertainties. As can be seen in Table 16, the compounds ATH, ZB, ALPI, PER, and DOPO scored high in at least one of the PBT categories. ATH and ZB exerted high toxicity to some species, while ALPI appeared to be persistent and may have moderate ecotoxicity, making them less suitable as alternative FRs. DOPO and MPP may be persistent, but this conclusion was based on fewer than two studies each, clearly indicating a lack of information. Most studies performed on PER showed that it had low persistence. Unfortunately, there is a lack of data on the bioaccumulation and in vitro toxicity of this compound. If future studies show that ALPI, DOPO, MPP, and PER are not bioaccumulative and toxic, they may still be considered as suitable FRs. Since two studies showed a moderate ecotoxicity for APP, it would not be a first choice alternative, although it scored low in all other PB&T categories. $Mg(OH)_2$, ZHS, and ZS did not show high bioaccumulation or toxicity, and so far, appear to be the most suitable HFFRs, but they also exhibited large data gaps, since none of the HFFRs were studied as elaborately as TPP. In Table 16, the HFFR are ranked according to suitability, with the highest PBT values on top. Future research is obviously necessary, to allow the PBT properties of each compound to be compared with those of the relevant halogenated flame retardant that it would replace. The different properties should be weighed and prioritized in a more extensive risk assessment, leading to a well-balanced trade-off between functionality and negative effects on humans and the environment.

9 Summary

Polymers are synthetic organic materials having a high carbon and hydrogen content, which make them readily combustible. Polymers have many indoor uses and their flammability makes them a fire hazard. Therefore, flame retardants (FRs) are incorporated into these materials as a safety measure. Brominated flame retardants (BFRs), which accounted for about 21% of the total world market of FRs, have several unintended negative effects on the environment and human health. Hence, there is growing interest in finding appropriate alternative halogen-free flame retardants (HFFRs). Many of these HFFRs are marketed already, although their environmental behavior and toxicological properties are often only known to a limited extent, and their potential impact on the environment cannot yet be properly assessed. Therefore, we undertook this review to make an inventory of the available data that exists (up to September 2011) on the physical–chemical properties, production volumes, persistence, bioaccumulation, and toxicity (PBT) of a selection of HFFRs that are potential replacements for BFRs in polymers.

Large data gaps were identified for the physical–chemical and the PBT properties of the reviewed HFFRs. Because these HFFRs are currently on the market, there is an urgent need to fill these data gaps. Enhanced transparency of methodology and data are needed to reevaluate certain test results that appear contradictory, and, if this does not provide new insights, further research should be performed. TPP has been studied quite extensively and it is clearly persistent, bioaccumulative, and toxic. So far, RDP and BDP have demonstrated low to high ecotoxicity and persistence. The compounds ATH and ZB exerted high toxicity to some species and ALPI appeared to be persistent and has low to moderate reported ecotoxicity. DOPO and MPP may be persistent, but this view is based merely on one or two studies, clearly indicating a lack of information. Many degradation studies have been performed on PER and show low persistence, with a few exceptions. Additionally, there is too little information on the bioaccumulation potential of PER. APP mostly has low PBT properties; however, moderate ecotoxicity was reported in two studies. $Mg(OH)_2$, ZHS, and ZS do not show such remarkably high bioaccumulation or toxicity, but large data gaps exist for these compounds also. Nevertheless, we consider the latter compounds to be the most promising among alternative HFFRs. To assess whether the presently reviewed HFFRs are truly suitable alternatives, each compound should be examined individually by comparing its PBT values with those of the relevant halogenated flame retardant. Until more data are available, it remains impossible to accurately evaluate the risk of each of these compounds, including the ones that are already extensively marketed.

Acknowledgments This research is part of the EU project ENFIRO (KP7-226563) and the financial support of the European Union is gratefully acknowledged.

References

ACD/Labs (2011) ACD/PhysChem Suite. Advanced Chemistry Development, Inc, Toronto, ON

Alaee M, Arias P, Sjodin A, Bergman A (2003) An overview of commercially used brominated flame retardants, their applications, their use patterns in different countries/regions and possible modes of release. Environ Int 29:683–689

Albemarle corporation (2003a) Material safety data sheet (MSDS)—Magnifin C grade flame retardants

Albemarle corporation (2003b) Material safety data sheet (MSDS)—Magnifin H grades flame retardants

Albemarle corporation (2003c) Material safety data sheet (MSDS)—Magnifin MV grades flame retardants

Albemarle Corporation (2009) Re: EPA-industry DecaBDE phase-out initiative. In: US Environmental Protection Agency (ed) Albemarle Corporation,Washington, DC. p 2

AluChem (2003) Material safety data sheet (MSDS)—magnesium hydroxide

Australian Government Regulator of Industrial Chemicals (1994) Flamtard H. In: Australian Department of Health and Ageing (administration) (ed). National industrial chemicals notification and assessment scheme NICNAS—full public reports

Australian Government Regulator of Industrial Chemicals (2000) Phosphoric acid, (1-methylethylidene) di-4,1-phenylene tetraphenyl ester (Fyrolflex BDP). In: Australian Department of Health and Ageing (administration) (ed) National industrial chemicals notification and assessment scheme NICNAS—full public reports

Australian Government Regulator of Industrial Chemicals (2005) Chemical in Exolit OP 1312. In: Australian Department of Health and Ageing (administration) (ed) National industrial chemicals notification and assessment scheme NICNAS—full public reports

Australian Government Regulator of Industrial Chemicals (2006) Melapur 200 and Polymer in Exolit OP 1312. In: Australian Department of Health and Ageing (administration) (ed) National industrial chemicals notification and assessment scheme NICNAS—full public reports

Baird DJ, Barber I, Bradley M, Calow P, Soares A (1989) The *Daphnia* bioassay—a critique. Hydrobiologia 188:403–406

Barceloux DG (1999) Zinc. J Toxicol Clin Toxicol 37:279–292

Bengtsson BE, Tarkpea M, Sletten T, Carlberg GE, Kringstad A, Renberg L (1986) Bioaccumulation and effects of some technical triaryl phosphate products in fish and Nitocra Spinipes. Environ Toxicol Chem 5:853–861

Bergh C, Torgrip R, Emenius G, Ostman C (2011) Organophosphate and phthalate esters in air and settled dust—a multi-location indoor study. Indoor Air 21:67–76

Berthon G (2002) Aluminium speciation in relation to aluminium bioavailability, metabolism and toxicity. Coord Chem Rev 228:319–341

Betts KS (2007) Formulating green flame retardants. Environ Sci Technol 41:7201–7202

Bilkei-Gorzo A (1993) Neurotoxic effect of enteral aluminium. Food Chem Toxicol 31:357–361

Bingham E, Cohrssen B, Powell CH (2001) Patty's toxicology, vol 1–9. Wiley, New York. p V6 967

Birnbaum LS, Staskal DF (2004) Brominated flame retardants: cause for concern? Environ Health Perspect 112:9–17

Boon JP, Lewis WE, Tjoen-A-Choy MR, Allchin CR, Law RJ, de Boer J, ten Hallers-Tjabbes CC, Zegers BN (2002) Levels of polybrominated diphenyl ether (PBDE) flame retardants in animals representing different trophic levels of the North Sea food web. Environ Sci Technol 36:4025–4032

Borax (2004) Material safety data sheet—Borogard ZB. Borax Inc.

Budenheim (2010) Human health and environmental fact sheet—melamine polyphosphate (Budit 3141 & Budit 3141 CA). Phosphorus, Inorganic & Nitrogen Flame Retardants Association, Pinfa

Chang TC, Wu KH, Wu TR, Chiu YS (1998) Thermogravimeter analysis study of a cyclic organophosphorus compound. Phosphorus, sulfur silicon. Relat Elem 139:45–56

Chemtura (2009) Re: DecaBDE phase-out initiative. In: U.S. Environmental Protection Agency (ed). Chemtura Corporation, Washington, DC

Chernysh EA, Bugerenk EF, Aksenov VI, Golubtso SA, Ponomare VV (1972) Organophosphorus heterocyclic-compounds. 3. Synthesis and conversion of 10-chloro-10-phospha-9-oxa-9, 10-dihydrophenanthrene. Zhurnal Obshchei Khimii 42:93–96

Ciba (2010) Human health and environmental fact sheet—melamine polyphosphate (Melapur 200). Phosphorus, Inorganic & Nitrogen Flame Retardants Association, Pinfa

Clariant (2007) Human health and environmental fact sheet—diethylphosphinic acid, aluminium salt (Depal, Exolit OP 1230, Exolit OP 930, Exolit OP 935 (R)). In: Clariant International Ltd (ed) Phosphorus, Inorganic & Nitrogen Flame Retardants Association, Pinfa

Clariant (2010) Human health and environmental fact sheet—ammonium polyphosphate (Exolit AP422). In: Clariant International Ltd (ed) Phosphorus, Inorganic & Nitrogen Flame Retardants Association, Pinfa

Clean Production Action, Rossi M, Heine L (2007) The green screen for safer chemicals: evaluating flame retardants for TV enclosures, vol Version 1.0, p 17

Cole JG, Mackay D (2000) Correlating environmental partitioning properties of organic compounds: the three solubility approach. Environ Toxicol Chem 19:265–270

Cummings JE, Kovacic JP (2009) The ubiquitous role of zinc in health and disease. J Vet Emerg Crit Care 19:215–240

Danish EPA, Lassen C, Løkk S, Andersen LI, Hansen E (1999) Brominated flame retardants—substance flow analysis and assessment of alternatives. p 227

Danish EPA, Stuer-Lauridsen F, Cohr K-H, Andersen TT (2007) Health and environmental assessment of alternatives to deca-BDE in electrical and electronic equipment. In: DHI Water & Environment (ed) Danish Environmental Protection Agency, EPA.

Danish EPA, Stuer-Lauridsen F, Havelund S, Birkved M (2000) Alternatives to brominated flame retardants—screening for environmental and health data. In: A/S C (ed) Danish Environmental Protection Agency, EPA

Dekant W (2009) Review of the toxicity of red phosphorus and related phosphorus based flame retardants. Institut für Toxikologie, Würzburg

Dimitrov SD, Dimitrova NC, Walker JD, Veith GD, Mekenyan OG (2002) Predicting bioconcentration factors of highly hydrophobic chemicals. Effects of molecular size. Pure Appl Chem 74:1823–1830

Dobry A, Keller R (1957) Vapor pressures of some phosphate and phosphonate esters. J Phys Chem 61:1448–1449

Eckert F, Klamt A (2010) COSMOtherm® Vers. C2.1 released 01.10. COSMOlogic GmbH & Co. KG

ENFIRO (2009) ENFIRO Webpage. www.enfiro.eu

ENFIRO partners, Leonards (Project Coordinator) PEG (2008) Life cycle assessment of environment-compatible flame retardants (prototypical case study)—project proposal—EU project (KP7-226563). In: ENFIRO (ed) p 72

Eto M, Hashimoto Y, Ozaki K, Kassai T, Sasaki Y (1975) Fungitoxicity and insecticide synergism of monothioquinol phosphate esters and related compounds. Bochu-Kagaku 40:160

European Chemicals Agency (ECHA) (1963-May-19) Pentaerythritol—experimental supporting study toxicity—toxicological information. ECHA

European Chemicals Agency (ECHA) Database (original study 1964) Pentaerythritol—acute toxicity; inhalation (toxicological information) experimental supporting study 1. ECHA

European Chemicals Agency (ECHA) Database (original study 1979-Feb-08) Pentaerythritol—biodegradation; in water (environmental fate and pathways) experimental supporting study 5. ECHA

European Chemicals Agency (ECHA) Database (original study 1980-Feb-29 & 1981) Pentaerythritol—biodegradation; in water (environmental fate and pathways) experimental study 6 WoE. ECHA

European Chemicals Agency (ECHA) Database (original study 1983) Pentaerythritol—aquatic toxicity; to microorganisms (ecotoxicological information) experimental study 2 weight of evidence (WoE). ECHA

European Chemicals Agency (ECHA) Database (original study 1985) Pentaerythritol—biodegradation; in water (environmental fate and pathways) experimental supporting study 2. ECHA

European Chemicals Agency (ECHA) Database (original study 1990-Feb-15) Pentaerythritol—acute toxicity; oral (toxicological information) experimental key study 1. ECHA

European Chemicals Agency (ECHA) Database (original study 1990-Sep-24) Pentaerythritol—biodegradation; in water (environmental fate and pathways) experimental supporting study 3. ECHA

European Chemicals Agency (ECHA) Database (original study 1991-Sep-16) Pentaerythritol—biodegradation; in water, screening (environmental fate and pathways) experimental key study 1. ECHA

European Chemicals Agency (ECHA) Database (original study 1992-May-30) Pentaerythritol—repeated dose toxicity; oral (toxicological information) experimental supporting study 2. ECHA

European Chemicals Agency (ECHA) Database (original study 1994-Dec-21) Pentaerythritol—genetic toxicity; in vitro (toxicological information) experimental key study 1. ECHA

European Chemicals Agency (ECHA) Database (original study 1994-Jan-25) Pentaerythritol—biodegradation; in water (environmental fate and pathways) experimental supporting study 4. ECHA

European Chemicals Agency (ECHA) Database (original study 1996-Jul-28) Pentaerythritol—repeated dose study; oral (toxicological information) experimental key study 1. ECHA

European Chemicals Agency (ECHA) Database (original study 1998a) Aluminium tris(dialkylphosphinate)—aquatic toxicity; short-term to aquatic invertebrates (ecotoxicological information) experimental key study 1. ECHA

European Chemicals Agency (ECHA) Database (original study 1998b) Aluminium tris(dialkylphosphinate)—aquatic toxicity; short-term to fish (ecotoxicological information) experimental key study 1. ECHA

European Chemicals Agency (ECHA) Database (original study 2005-Apr-13) Aluminium tris(dialkylphosphinate)—aquatic toxicity; long-term to aquatic invertebrates (ecotoxicological information) experimental key study 1. ECHA

European Chemicals Agency (ECHA) Database (original study 2008) Aluminium tris(dialkylphosphinate)—genetic toxicity; in vivo (toxicological information) experimental key study 1. ECHA

European Chemicals Agency (ECHA) Database (Accessed 2011) Triphenylphosphate—classification and labelling; GHS & DSD—DPD ECHA

European Chemicals Agency (ECHA) Database (original study 1964b) Pentaerythritol—acute toxicity; oral (toxicological information) experimental supporting study 2. ECHA

European Chemicals Agency (ECHA) Database (original study 1993-Jan-03) Pentaerythritol—aquatic toxicity; short-term to fish (ecotoxicological information) experimental key study 1. ECHA

European Chemicals Agency (ECHA) Database (original study 1993-March-01a) Pentaerythritol—aquatic toxicity; long-term to aquatic invertebrates (ecotoxicological information) experimental key study 1. ECHA

European Chemicals Agency (ECHA) Database (original study 1993-March-01b) Pentaerythritol—aquatic toxicity; short-term to aquatic invertebrates (ecotoxicological information) experimental key study 1. ECHA

European Chemicals Agency (ECHA) Database (original study 1994-Dec-07) Pentaerythritol—genetic toxicity; in vitro (toxicological information) experimental key study 3. ECHA

European Chemicals Agency (ECHA) Database (original study 1996-Jul-25) Pentaerythritol—acute toxicity; oral (toxicological information) experimental supporting study 4. ECHA

European Chemicals Agency (ECHA) Database (original study 1998a) Aluminium tris(dialkylphosphinate)—acute toxicity; dermal (toxicological information) experimental key study 1. ECHA

European Chemicals Agency (ECHA) Database (original study 1998b) Aluminium tris(dialkylphosphinate)—acute toxicity; oral (toxicological information) experimental key study 1. ECHA

European Chemicals Agency (ECHA) Database (original study 1998c) Aluminium tris(dialkylphosphinate)—aquatic toxicity; to aquatic algae and cyanobacteria (ecotoxicological information) experimental key study 1. ECHA

European Chemicals Agency (ECHA) Database (original study 1998d) Aluminium tris(dialkylphosphinate)—aquatic toxicity; to microorganisms (ecotoxicological information) experimental key study 1. ECHA

European Chemicals Agency (ECHA) Database (original study 1998e) Aluminium tris(dialkylphosphinate)—genetic toxicity; in vitro (toxicological information) experimental key study 1. ECHA

European Chemicals Agency (ECHA) Database (original study 1998f) Aluminium tris(dialkylphosphinate)—genetic toxicity; in vitro (toxicological information) experimental key study 2. ECHA

European Chemicals Agency (ECHA) Database (original study 1998 g) Aluminium tris(dialkylphosphinate)—repeated dose toxicity; oral (toxicological information) experimental key study 1. ECHA

European Chemicals Agency (ECHA) Database (original study 2008) Aluminium tris(dialkylphosphinate)—toxicity to reproduction (toxicological information) experimental supporting study 1. ECHA

European Chemicals Agency (ECHA) Database (original study 2009a) Aluminium tris(dialkylphosphinate)—aquatic toxicity; long-term to fish (ecotoxicological information) experimental key study 1. ECHA

European Chemicals Agency (ECHA) Database (original study 2009b) Aluminium tris(dialkylphosphinate)—biodegradation; in water (environmental fate and pathways) experimental key study 1. ECHA

European Chemicals Agency (ECHA) Database (original study 2009c) Aluminium tris(dialkylphosphinate)—biodegradation; in water (environmental fate and pathways) experimental key study 2. ECHA

European Chemicals Agency (ECHA) Database (original study 2009d) Pentaerythritol—log Kow (physical chemical properties) experimental key study 1. ECHA

European Chemicals Agency (ECHA) Database (original study 2009e) Pentaerythritol—Meltingpoint (Physical Chemical Properties) Experimental Key Study 1. ECHA

European Chemicals Agency (ECHA) Database (original study 2009f) Pentaerythritol—water solubility (physical chemical properties) experimental key study 1. ECHA

European Chemicals Agency (ECHA) Database (original study 2010-April-29) Pentaerythritol—aquatic toxicity; to microorganisms (ecotoxicological information) experimental key study 1. ECHA

European Chemicals Agency (ECHA) Database (original study 2010-Aug-2) Pentaerythritol—stability; hydrolosis (environmental fate and pathways) experimental study 1 WoE. ECHA

European Chemicals Agency (ECHA) Database (original study 2010-Jan-31) Pentaerythritol—vapour pressure (physical chemical properties) calculated key study 1. ECHA

European Chemicals Agency (ECHA) Database (original study 2010-Okt-2) Pentaerythritol—genetic toxicity; in vitro (toxicological information) experimental key study 2. ECHA

European Chemicals Bureau (1998 (2005 updated)) IUCLID dataset pentaerythritol (CAS No. 115-77-5), SIDS initial assessment report for 8th SIAM. France, 28–30 October 1998. In: United Nations Environment Programme (UNEP) Publications (ed). United Nations Environment Programme, Organisation for Economic Co-operation and Development (OECD) Screening Information DataSets (SIDS)

European Chemicals Bureau (2000a) IUCLID Dataset aluminium trihydroxide (CAS: 21645-51-2). In :European Chemicals Bureau (ed) European Chemicals Bureau, Institute of Health and Consumer Protection, Joint Research Centre, JRC, European Commission, Brussel, Belgium

European Chemicals Bureau (2000b) IUCLID dataset boric acid (CAS: 10043-35-3). In: European Chemicals Bureau (ed) European Chemicals Bureau, Institute of Health and Consumer Protection, Joint Research Centre, JRC, European Commission, Brussel, Belgium

European Chemicals Bureau (2000c) IUCLID dataset magnesium hydroxide (CAS: 1309-42-8). In: European Chemicals Bureau (ed) European Chemicals Bureau, Institute of Health and Consumer Protection, Joint Research Centre, JRC, European Commission, Brussel, Belgium

European Chemicals Bureau (2000d) IUCLID dataset of polyphosphoric acids, ammonium salts (APP) (CAS: 68333-79-9). In: European Chemicals Bureau (ed)European Chemicals Bureau, Institute of Health and Consumer Protection, Joint Research Centre, JRC, European Commission, Brussel, Belgium

European Chemicals Bureau (2000e) IUCLID dataset pentaerythritol (CAS: 115-77-5). In: European Chemicals Bureau (ed) European Chemicals Bureau, Institute of Health and Consumer Protection, Joint Research Centre, JRC, European Commission, Brussel, Belgium

European Chemicals Bureau (2000f) IUCLID dataset triphenyl phosphate (CAS: 115-86-6). In: European Chemicals Bureau (ed) European Chemicals Bureau, Institute of Health and Consumer Protection, Joint Research Centre, JRC, European Commission, Brussel, Belgium

European Chemicals Bureau (2002) IUCLID dataset triphenyl phosphate (CAS No. 115-86-6), SIDS Initial Assessment Report for SIAM 15. Boston, MA, 22–25 October 2002. In: United Nations Environment Programme (UNEP) Publications (ed): United Nations Environment Programme, Organisation for Economic Co-operation and Development (OECD) Screening Information DataSets (SIDS)

European Chemicals Bureau (2011) European Chemical Substance Information System (ESIS). European Chemicals Bureau, Joint Research Centre, JRC, European Commission

European Chemicals Bureau, Pakalin S, Cole T, Steinkellner J, Nicolas R, Tissier C, Munn S, Eisenreich S (2007) Review on production processes of decabromodiphenyl ether (decaBDE) used in Polymeric applications in electrical and electronic equipment, and assessment of the availability of potential alternatives to DecaBDE. In: Europen Chemicals Bureau (ed) European Chemicals Bureau, Institute of Health and Consumer Protection, Joint Research Centre, JRC, European Commission, Brussel, Belgium

European Flame Retardants Association (EFRA), Cefic (2006) Zinc borate flame retardants fact sheet In: Cefic European Chemical Industry Council (ed). Cefic European Chemical Industry Council

European Parliament (E.P.) (2002) Report A5-0437/2002

European Union (1993) Council Regulation No 793/93/EEC of 23 March 1993 on the evaluation and control of risks of existing substances. (L84/1). In: European Parliament and the Council on the European Union (ed) European Union, p 75

European Union (2006) Regulation (EC) No 1907/2006 of the European Parliament and of the council concerning the Registration, Evaluation, Authorisation and Restriction of Chemicals (REACH), establishing a European Chemicals Agency. In: European Parliament and the Council on the European Union (ed), European Union, p 849

European Union (2008) Regulation (EC) No 1272/2008 of the European Parliament and of the council on classification, labelling and packaging of substances and mixtures. In: European Parliament and the Council on the European Union (ed), European Union, p 1355

Fängström B, Strid A, Grandjean P, Weihe P, Bergman Å (2005) A retrospective study of PBDEs and PCBs in human milk from the Faroe Islands. Environmental Health: A Global Access Science Source 4, 12

Fisher Scientific (1999 (2008 updated)) Material safety data sheet (MSDS)—magnesium hydroxide (13405). Fisher Scientific

Flaskos J, McLean WG, Hargreaves AJ (1994) The toxicity of organophosphate compounds towards cultured PC12 Cells. Toxicol Lett 70:71–76

Fosmire GJ (1990) Zinc toxicity. Am J Clin Nutr 51:225–227

Gant DB, Eldefrawi ME, Eldefrawi AT (1987) Action of organophosphates on GABAA receptor and voltage-dependent chloride channels. Fundam Appl Toxicol 9:698–704

Gardner JR (1988a) Acute oral toxicity to rats of zinc hydroxystannate (ZHS). Huntingdon Research Centre (HRC), Huntingdon

Gardner JR (1988b) Acute oral toxicity to rats of zinc stannate (ZS). Huntingdon Research Centre (HRC), Huntingdon

Gelest (2008) Material safety data sheet of zinc stannate—SNZ9760. Gelest, Inc.

German Federal Environmental Agency, Leisewitz A, Kruse H, Schramm E (2001) Substituting environmentally relevant flame retardants: assessment fundamentals—results and summary overview. In: Environmental research plan of the German Federal Ministry for the Environment, Nature Conservation and Nuclear Safety, Contractor: Öko-Recherche. Büro für Umweltforschung und-beratung GmbH, Frankfurt am Main, p 204

Gobas FAPC, Kelly BC, Arnot JA (2003) Quantitative structure activity relationships for predicting the bioaccumulation of POPs in terrestrial food-webs. Qsar Comb Sci 22:329–336

Hachiya N (1987) Evaluation of chemical genotoxicity by a series of short term tests. Akita Igaku 14:269–292

Henrich R, Ryan BM, Selby R, Garthwaite S, Morrissey R, Freudenthal RI (2000) Two-generation oral (diet) reproductive toxicity study of resorcinol bis-diphenylphosphate (Fyrolflex RDP) in rats. Int J Toxicol 19:243–255

Hilal SH, Karickhoff SW, Carreira LA (2003) Sparc On-line Calculator 4.5—predicting melting point, vapour pressure & Henry's law constant—based on "Prediction of the vapor pressure boiling point, heat of vaporization and diffusion coefficient of organic compounds". Qsar Comb Sci 22:565–574

Hilal SH, Karickhoff SW, Carreira LA (2004) Sparc On-line Calculator 4.5—predicting water solubility & log Kow—based on "Prediction of the solubility, activity coefficient and liquid/liquid partition coefficient of organic compounds". Qsar Comb Sci 23:709–720

Hites RA (2004) Polybrominated diphenyl ethers in the environment and in people: a meta-analysis of concentrations. Environ Sci Technol 38:945–956

Hoenicke R, Oros DR, Oram JJ, Taberski KM (2007) Adapting an ambient monitoring program to the challenge of managing emerging pollutants in the San Francisco Estuary. Environ Res 105:132–144

Hubbard CM, Redpath GT, Macdonald TL, VandenBerg SR (1989) Modulatory effects of aluminum, calcium, lithium, magnesium, and zinc oons on [3H]MK-801 binding in human cerebral cortex. Brain Res 486:170–174

ICL (2009) Commitment letter to EPA, Re: voluntary phase out of DecaBDE.In: U.S. Environmental Protection Agency (ed). ICL Industrial Products, Washington, DC. p 2

ICL Industrial Products (2011) Material safety data sheet (MSDS)—Fyrolflex RDP.

Illinois EPA (2007) Report on alternatives to the flame retardant DecaBDE: evaluation of toxicity, availability, affordability, and fire safety issues—a report to the Governor and the General Assembly. p 86

ITRI (2009) Technical bulletin—zinc stannates and zinc hydroxy stannate. ITRI technical bulletin, vols 2 & 3

Joseph Storey & Co. Ltd. (1994) Material safety data sheet (MSDS)—Storflam ZHS (zinc hydroxy stannate). Joseph Storey & Co. Ltd, Lancaster

Knie VJ, Hälke A, Juhnke I, Schiller W (1983) Ergebnisse der Untersuchungen von chemischen Stoffen mit vier Biotests [Results of studies on chemical substances with four biotests]. Deutsche Gewässerkundliche Mitteilungen 27:77–79

Kucera T, Horakova H, Sonska A (2008) Toxic metal ions in photoautotrophic organisms. Photosynthetica 46:481–489

Kuo Ching Chemical Co Ltd. (2009) Specification Sheet—KFR DOPO. Kuo Ching Chemical Co, Ltd

Lewis RL (2000) Sr. Sax's dangerous properties of industrial materials, 10th edn. Wiley, New York

Maine Department of Environmental Protection, Maine Center for Disease Control & Prevention (2007) Brominated flame retardants: third annual report to the Maine Legislature. Maine, United States of America

MaKuang Chemical Co. Ltd. (2009) Data sheet—MK 68 (DOPO). MaKuang Chemical Co, Ltd. Taichung, Taiwan

Maret W, Sandstead HH (2006) Zinc requirements and the risks and benefits of zinc supplementation. J Trace Elem Med Biol 20:3–18

Martin Marietta Magnesia Specialties LLC (MMMS), Walter MD, Wajer MT (2010) Overview of flame retardants including magnesium hydroxide. Martin Marietta Magnesia Specialties LLC, Baltimore, MA

Martinez CE, Motto HL (2000) Solubility of lead, zinc and copper added to mineral soils. Environ Pollut 107:153–158

Matsukami H, Honda M, Nakamura A, Takasuga T (2010) Analysis of monomeric and oligomeric phosphate esters in indoor air and house dust by GC-MS and LC-MS. Extended abstract, fifth International Workshop on brominated flame retardants BFR 2010

Mayer FL, Adams WJ, Finley MT, Michael PR, Mehrle PM, Saeger VW (1981) Phosphate ester hydraulic fluids: an aquatic environmental assessment of Pydrauls 50E and 115E. In: Branson DR, Dickson KL (eds) Aquatic toxicology and hazard assessment, 4th conference, vol STP 737. American Society for Testing and Materials (ASTM): Philadelphia, PA. pp 103–123

McDonald SF, Hamilton SJ, Buhl KJ, Heisinger JF (1996) Acute toxicity of fire control chemicals to Daphnia magna (Straus) and Selenastrum capricornutum (Printz). Ecotoxicol Environ Saf 33:62–72

McEntee TE (1987) PC-Nomograph-Programs to enhance PC-GEMS estimates of physical properties for organic chemicals Version 2.0—EGA/CGA. MS DOS: 1987/12/04. The Mitre Corporation, MSDOS

McPherson A, Thorpe B, Blake A (2004) Brominated flame retardants in dust on computers: the case for safer chemicals and better computer design. Clean Protection Action. p 43

Meerts I, Letcher RJ, Hoving S, Marsh G, Bergman A, Lemmen JG, van der Burg B, Brouwer A (2001) In vitro estrogenicity of polybrominated diphenyl ethers, hydroxylated PBDEs, and polybrominated bisphenol A compounds. Environ Health Perspect 109:399–407

Merck & Co. Inc. (2001) The Merck Index—13th edn—triphenyl phosphate darmstadt. Merck KGaA, Darmstadt, Duitsland

Merck & Co. Inc. (2006) The Merck Index—14th edn—triphenyl phosphate (Monograph nr: 09742). Merck KGaA, Darmstadt, Duitsland

Merck Chemicals—Product Information (Merck Website) Magnesiumhydroxide 105870 (CAS 1309-42-8). http://www.merck-chemicals.nl/magnesiumhydroxide/MDA_CHEM-105870/p_uuid. Merck KGaA, Darmstadt, Duitsland

Merck Chemicals—Product Information (Merck Website) Pentaerythritol 807331 (CAS 115-77-5). http://www.merck-chemicals.nl/pentaerythritol/MDA_CHEM-807331/p_uuid. Merck KGaA, Darmstadt, Duitsland

Merck Chemicals—Product Information (Merck Website) Trifenyl Fosfaat 821197 (CAS 115-86-6). http://www.merck-chemicals.nl/trifenyl-fosfaat/MDA_CHEM-821197/p_uuid. Merck KGaA, Darmstadt, Duitsland

Meylan WM, Howard PH (1991) Bond contribution methods for estimating Henry's law constants. Environ Toxicol Chem 10:1283–1293

Meylan WM, Howard PH (1995) Atom fragment contribution methods for estimating octanol-water partition-coefficients. J Pharm Sci 84:83–92

Meylan WM, Howard PH, Boethling RS (1996) Improved method for estimating water solubility from octanol water partition coefficient. Environ Toxicol Chem 15:100–106

Mortelmans K, Zeiger E (2000) The Ames Salmonella/microsome mutagenicity assay. Mut Res-Fund Mol Mech Mutagen 455:29–60

Muir DCG, Yarechewski AL, Grift NP (1989) Biodegradation of 4 triaryl alkyl phophate-esters in sediment under various temperature and redox conditions. Toxicol Environ Chem 18:269–286

Nabaltec (2009) Human health and environmental fact sheet—magnesium hydroxide (Brucite) (Apymag). Phosphorus, Inorganic & Nitrogen Flame Retardants Association, Pinfa.

Nagajyoti PC, Lee KD, Sreekanth TVM (2010) Heavy metals, occurrence and toxicity for plants: a review. Environ Chem Lett 8:199–216

Neely B, Blau GE (1985) Environmental exposure from chemicals. CRS, Boca Ratón, FL

Nordin H (2007) Human health and environmental fact sheet—Melapur 200 70. Phosphorus, Inorganic & Nitrogen Flame Retardants Association, Pinfa. p 2

Norén K, Meironyté D (2000) Certain organochlorine and organobromine contaminants in Swedish human milk in perspective of past 20–30 years. Chemosphere 40:1111–1123

O'Connell S, Whitley A, Brady T, Ching S, Fong A, Bergman R, Burkitt J (2004) Environmental assessment of halogen-free printed circuit boards, a design for environment (DfE) project with the high density packaging user group (HDPUG). Computer Society Washington, Washington, DC

Organisation for Economic Co-operation and Development (OECD) (1992) OECD guideline for testing of chemicals—ready biodegradability 301. In: OECD/OCDE 301, vol Section 3. Degradation and accumulation: OECD.

OSPAR (2001 (2004 updated)) Certain brominated flame retardants – polybrominated diphenyle-thers, polybrominated biphenyls, hexabromo cyclododecane, vol OSPAR Priority Substances Series In: The Convention for the Protection of the marine Environment of the North-East Atlantic (ed)

Padilla SS, Grizzle TB, Lyerly D (1987) Triphenyl phosphite: in vivo and in vitro inhibition of rat neurotoxic esterase. Toxicol Appl Pharmacol 87:249–256

Perstorp Specialty Chemicals AB (2008) Material safety data sheet of pentaerythritol mono, pentaerythritol nitration. Perstorp Specialty Chemicals AB

Peterson HG, Healey FP, Wagemann R (1984) Metal toxicity to algae—a highly pH dependent phenomenon. Can J Fish Aquat Sci 41:974–979

Rio Tinto Alcan (RTA) (2008a) Material safety data sheet (MSDS)—alumina hydrate (000219). Rio Tinto Alcan

Rio Tinto Alcan (RTA) (2008b) Material safety data sheet (MSDS)—Flamtard H (000042) zinc hydroxy stannate. Rio Tinto Alcan (RTA)

Rio Tinto Alcan (RTA) (2008c) Material safety data sheet (MSDS)—Flamtard S (000038) zinc stannate. Rio Tinto Alcan (RTA)

Robrock KR, Korytar P, Alvarez-Cohen L (2008) Pathways for the anaerobic microbial debromination of polybrominated diphenyl ethers. Environ Sci Technol 42:2845–2852

Rout GR, Das P (2003) Effect of metal toxicity on plant growth and metabolism: I. Zinc. Agronomie 23:3–11

Ryan BM, Henrich R, Mallett E, Freudenthal RI (2000) Developmental toxicity study of orally administered resorcinol bis-diphenylphosphate (RDP) in rabbits. Int J Toxicol 19:257–264

Saeger VW, Hicks O, Kaley RG, Michael PR, Mieure JP, Tucker ES (1979) Environmental fate of selected phosphate-esters. Environ Sci Technol 13:840–844

Sasaki K, Takeda M, Uchiyama M (1981) Toxicity, absorption and elimination of phosphoric acid triesters by killifish and goldfish. Bull Environ Contam Toxicol 27:775–782

Schantz SL, Widholm JJ, Rice DC (2003) Effects of PCB exposure on neuropsychological function in children. Environ Health Perspect 111:357–376

Schecter A, Harris TR, Brummitt S, Shah N, Paepke O (2008) PBDE and HBCD brominated flame retardants in the USA, update 2008: levels in human milk and blood, food, and environmental samples. Epidemiology 19:S76–S76

Schenker U, MacLeod M, Scheringer M, Hungerbuhler K (2005) Improving data quality for environmental fate models: a least-squares adjustment procedure for harmonizing physicochemical properties of organic compounds. Environ Sci Technol 39:8434–8441

Shimizu H, Suzuki Y, Takemura N, Goto S, Matsushita H (1985) The results of microbial mutation test for forty-three industrial chemicals. Sangyo Igaku Jpn J Ind Health 27:400–419

Shüürmann G, Ebert R-U, Nendza M, Dearden JC, Paschke A, Kühne R (2007) Chapter 9: predicting fate-related physicochemical properties. In: Leeuwen CJV, Vermeire TG (eds) Risk assessment of chemicals: an introduction. Springer, Heidelberg, pp 379–382

Smith MI, Elvove E, Frazier WH (1930) The pharmacological action of certain phenol esters with special reference to the etiology of so-called ginger paralysis. Public Health Reports 45:2509–2524

Smith MI, Engel EW, Stohlman EF (1932) Further studies on the pharmacology of certain phenol esters with special reference to the relation of chemical cestitution and physiologic action. National Institute of Health Bulletin 160:1–53

Snyder R (ed) (1990) Nitrogen and phosphorus solvents. Ethyl browning's toxicity and metabolism of industrial solvents, vol II. Elsevier, Amsterdam, p 491

Spurgeon DJ, Lofts S, Hankard PK, Toal M, McLellan D, Fishwick S, Svendsen C (2006) Effect of pH on metal speciation and resulting metal uptake and toxicity for earthworms. Environ Toxicol Chem 25:788–796

SRI Consulting (SRIC) (2004) Worldwide FR consumption and geographical distribution. Zürich, Switzerland

Stevens GC, Mann AH (1999) Risks and benefits in the use of flame retardants in consumer products. In: For the UK Department of Trade and Industry (ed) Polymer Reserch Centre, University of Surrey, Guildford, Surrey. p 197

Sundkvist AM, Olofsson U, Haglund P (2010) Organophosphorus flame retardants and plasticizers in marine and fresh water biota and in human milk. J Environ Monit 12:943–951

Supresta (2006) Material saftey data sheet (MSDS) — Fyrolflex BDP. Product Safety Department, Supresta

Syracuse Research Corporation (SRC) (2006) An assessment of potential health and environmental impacts of RDP and BAPP, two phosphate-based alternatives to Deca-BDE for use in electronics. Conducted for the Washington State Departments of Ecology and Health

Syracuse Research Corporation (SRC) (2009) SRC physical property database (PHYSPROP, Pentaerythritol: 115.77.5). SRC Inc.

Syracuse Research Corporation (SRC) (2011) Toxic Substance Control Act Test Submission Database (TSCATS). SRC Inc.

The Subcommittee on Flame-Retardant Chemicals CoT, Board on Environmental Studies and Toxicology, National Research Council (2000) Toxicological risks of selected flame-retardant chemicals. National Academy Press, Washington, DC

U.K. Environment Agency, Brooke DN, Crookes MJ, Burns J, Quarterman P (2009a) Environmental risk evaluation report: tetraphenyl resorcinol diphosphate (CAS no. 57583-54-7). SCHO0809BQUL-E-P, Bristol, UK. p 78

U.K. Environment Agency, Brooke DN, Crookes MJ, Quarterman P, Burns J (2009b) Environmental risk assessment report: summary and overview aryl phosphate esters (Draft). Bristol, UK

U.K. Environment Agency, Fisk PR, Girling AE, Wildey RJ (2003) Prioritisation of flame retardants for environmental risk assessment. p 129

U.S. Department of Health and Human Services, Public Health Service, Agency for Toxic Substances and Disease Registry (2009) Toxicological profile for phosphate ester flame retardants (Draft)

U.S. Environmental Protection Agency (EPA) (2008) Flame retardants in printed circuit boards (Draft). In: Design for the Environment (DfE) Flame Retardant in Printed Circuit Board Partnership (ed). p 273

U.S. EPA (2005) Environmental profiles of chemical flame-retardant alternatives for low-density polyurethane foam—Volume 1. In: Furniture Flame Retardancy Partnership (ed). p 153

U.S. EPA (2006) Non-confidential 2006 IUR records by chemical. Inventory Update Reporting (IUR). United States Environmental Protection Agency, Washington, DC

U.S. EPA (2008) Flame retardants in printed circuit boards. In: Design for the Environment (DfE) Flame Retardant in Printed Circuit Board Partnership (ed). p 273

U.S. EPA (2011) Estimation Programs Interface Suite™ for Microsoft® Windows v4.10. United States Environmental Protection Agency, Washington, DC

U.S. EPA (2012) Ecotox Database. vol Version 4. U.S. Environmental Protection Agency

UNEP OECD SIDS (1998, 2005) Pentaerythritol (CAS No. 115-77-5), SIDS Initial assessment report for 8th SIAM. France, 28–30 October, 1998. In: United Nations Environment Programme Publications (ed): United Nations Environment Programme, Organisation for Economic Co-operation and Development (OECD) Screening Information DataSets (SIDS)

UNEP OECD SIDS (2002a) Triphenyl phosphate (CAS No. 115-86-6). In: United Nations Environment Programme Publications (ed): United Nations Environment Programme, Organisation for Economic Co-operation and Development (OECD) Screening Information DataSets (SIDS)

UNEP OECD SIDS (2002b) Triphenyl phosphate CAS No. 115-86-6. In: United Nations Environment Programme Publications (ed): United Nations Environment Programme, Organisation for Economic Co-operation and Development (OECD) Screening Information DataSets (SIDS)

UNEP OECD SIDS (2007) Ammonium polyphosphate APP (CAS No. 68333-79-9). In: United Nations Environment Programme Publications (ed): United Nations Environment Programme, Organisation for Economic Co-operation and Development (OECD) Screening Information DataSets (SIDS)

United Nations Environment Programme (UNEP) (2008) Guidance on flame-retardant alternatives to pentabromodiphenyl ether (PentaBDE). The Stockholm Convention on Persistent Organic Pollutants of the United Nations Environment Programme (UNEP), Stockholm, Sweden.

United Nations Environment Programme (UNEP), Stockholm Convention—Press release (8 May 2009) Governments unite to step-up reduction on global DDT reliance and add nine new chemicals under international treaty. Geneva

Vainiotalo S, Verkkala E, Savolainen H, Nickels J, Zitting A (1987) Acute biological effects of commercial cresyl diphenyl phosphate in rats. Toxicology 44:31–44

Washington State Department of Ecology and Department of Health (2006) Washington State polybrominated diphenyl ether (PBDE) chemical action plan: final plan. Department of Ecology Publication No. 05-07-048, Department of Health Publication No. 334–079, Olympia, WA. p 307

William Blythe (2010a) Material safety data sheet (MSDS)—Flamtard H (zinc hydroxy stannate). Phosphorus, Inorganic & Nitrogen Flame Retardants Association, Pinfa

William Blythe (2010b) Material safety data sheet (MSDS)—Flamtard S (zinc stannate). Phosphorus, Inorganic & Nitrogen Flame Retardants Association, Pinfa

Wills JH, Barron K, Groblewski GE, Benitz KF, Johnson MK (1979) Does triphenyl phosphate produce delayed neurotoxic effects? Toxicol Lett 4:21–24

World Health Organisation (WHO) (1994) Polybrominated biphenyls. Environmental Health Criteria-152, vol International Program on Chemical Safety, Geneva, Switzerland

World Health Organization (WHO) (1991) Triphenyl phosphate. Environmental Health Criteria-111, vol International Program on Chemical Safety, Geneva, Switzerland

World Health Organization (WHO) (1997) Flame retardants: a general introduction. Environmental Health Criteria-192, vol International Program on Chemical Safety, Geneva, Switzerland

Wu J-P, Luo X-J, Zhang Y, Luo Y, Chen S-J, Mai B-X, Yang Z-Y (2008) Bioaccumulation of polybrominated diphenyl ethers (PBDEs) and polychlorinated biphenyls (PCBs) in wild aquatic species from an electronic waste (e-waste) recycling site in South China. Environ Int 34:1109–1113

Zatta P, Perazzolo M, Facci L, Skaper SD, Corain B, Favarato M (1992) Effects of aluminum speciation on murine neuroblastoma cells. Mol Chem Neuropathol 16:11–22

Zeiger E, Anderson B, Haworth S, Lawlor T, Mortelmans K, Speck W (1988) Salmonella mutagenicity tests: III. Results from the testing of 255 chemicals. Environ Mutagen 11:158

Radiation Exposure and Adverse Health Effects of Interventional Cardiology Staff

Chandrasekharan Nair Kesavachandran, Frank Haamann, and Albert Nienhaus

Contents

C.N. Kesavachandran (✉)
Centre for Epidemiology and Health Services Research in the Nursing Profession (CV care),
University Medical Centre Hamburg-Eppendorf, Martinistraße 52, Hamburg 20246, Germany

Epidemiology Division, CSIR-Indian Institute of Toxicology Research,
Lucknow, Uttar Pradesh 226001, India
e-mail: ckesavac@uke.uni-hamburg.de; ckesavachandran@gmail.com

F. Haamann
Department of Occupational Health Research, Institute for Statutory Accident Insurance
and Prevention in the Health and Welfare Services, Pappelallee 33/35/37,
Hamburg 22089, Germany

A. Nienhaus
Centre for Epidemiology and Health Services Research in the Nursing Profession (CV care),
University Medical Centre Hamburg-Eppendorf, Martinistraße 52, Hamburg 20246, Germany

Department of Occupational Health Research, Institute for Statutory Accident Insurance
and Prevention in the Health and Welfare Services, Pappelallee 33/35/37,
Hamburg 22089, Germany

D.M. Whitacre (ed.), *Reviews of Environmental Contamination and Toxicology*,   73
Reviews of Environmental Contamination and Toxicology 222,
DOI 10.1007/978-1-4614-4717-7_2, © Springer Science+Business Media New York 2013

1 Introduction

Coronary angiography (CA), percutaneous coronary intervention (PCI), catheter-based structural heart intervention, electrophysiological studies, and arrhythmia ablation are procedures that help cardiologists ensure better clinical diagnosis and treatment (Dawkins et al. 2005). During these procedures, catheters, guide wires, and other devices are visualized and guided by using real-time fluoroscopy. Therefore, operators are inevitably exposed to radiation (Kim and Miller 2009). Compared to other departments (radiology, urology, operating rooms, etc.), the cardiovascular or catheterization laboratory is generally considered to be an area of high radiation exposure (Raza 2011). Interventional cardiology (IC) staff is exposed more radiation per year than are radiologists by a factor of two to three (Picano et al. 2007). Invasive cardiology procedures have increased tenfold in the past decade, and growth in the field has been accompanied by concern for the safety of such staff (Picano et al. 2007).

Junior cardiologists are exposed to 60% more radiation than are their seniors (Watson et al. 1997). This difference is largely a result of younger staff taking longer duration to position fluoroscopic catheters, due to their lesser skill and shorter practice (Kottou et al. 2001). High workloads, the complexity of procedures and the lack of IC specialists in hospitals are growing concerns in the health sector (ICRP 2000; Vano et al. 1998a, b). The practices employed in catheterization laboratory facilities have become routine not only in Western societies but also in the Asia Pacific region (Rotter et al. 2003; Asian Network of cardiologists 2007; Tsapaki et al. 2011).

The staff who work in IC departments employs relatively high amounts of radiation (Delichas et al. 2003), and face the risk of developing cataracts after several years of work exposure, if radiation protection tools are not properly used (Sim et al. 2010). Cumulative X-ray doses imposed on the lenses of IC staff's eyes are often high (Vano et al. 2010a). Radiation-induced cataracts are distinct from naturally occurring ones, because they form in the posterior pole of the lens (Vano et al. 2010a). The increased incidence of lenticular changes that are occurring in IC staff, and its association with radiation doses is an important finding that underlines the need to address current concerns about the threshold dose for cataract formation (Bjelac et al. 2010).

During procedures, IC staff members are directly exposed to radiation that is reflected (scattered) from the patient (primary) and to a lesser extent from the walls of the room (secondary) (Maeder et al. 2005). The imposition of radiation dose limits for unprotected parts of the body, like eyes, hands, and the thyroid gland is crucial among IC staff, if they are to avoid the development of cataracts, cancers of the brain, skin, or thyroid (Raza 2011; Finkelstein 1998). The Ionising Radiation Regulation, introduced in 1999, reduced the maximum whole body dose for exposed personnel, but did not revise the maximum dose to the extremities (Hafez et al. 2005). The current annual dose limit is 20 mSv for the body, 150 mSv for the thyroid or eyes, and 500 mSv for the hands (International guidelines, ICRP). The recommended occupational dose of radiation for medical staff in Germany is 500 mSv for hands, 150 mSv for eyes, and 300 mSv for the thyroid [German Guidelines 2003].

In this systematic review, we address the following research questions:

1. Are radiation doses for IC staff within the prescribed limits?
2. Do current exposure levels produce adverse health effects for IC staff?
3. Are protective measures taken against radiation exposure adequate?

2 Criteria Applied to This Systematic Review

We performed a systematic literature search by using PubMed and EMBASE and by inputting appropriate keywords into the Google search engine. The keywords used were "radiation dose," combined with "dose," "interventional," "cardiologists," "technical," "nurses," "hands," "fingers," "neck," "thyroid," "eyes," "forehead," "health," "effects," "cataract," and "cancer." The search entailed the period from January 1990 to October 2011. We also searched through reference lists of selected prominent studies for relevant publications. However, no additional eligible publication was identified by searching through these reference lists. The literature search was conducted using the Preferred Reporting Items for Systematic Reviews and Meta-Analyses (PRISMA) guidelines. The process for identifying, screening, determining eligibility, and inclusion of databases for this study is shown in a flow chart (Fig. 1). The flow chart was developed from the PRISMA flow diagram that is used for reporting databases in systematic reviews (Moher et al. 2009). The review protocol for PRISMA was based on the information given at the following website: http://www.prisma-statement.org/statement.htm.

In the present review, we included publications that addressed the types of topical information showed below.

Study design: Cohort or cross-sectional studies.
Study population: Interventional cardiology staff.
Exposure: Annual and per procedure radiation dose in the workplace for different anatomical locations (hands/fingers, eyes/forehead, neck/thyroid).
Languages: German and English.

We assessed the methodological quality of the literature and classed studies as being "moderate" or "good." A study was deemed to be of "moderate" quality, if it did not include dosimetric measurements of the eyes, thyroid gland and hands of IC staff. A study was rated as "good," if the radiation dose of IC staff for these anatomical locations (eyes, thyroid, and hands) was measured and/or adverse health concerns related to these findings discussed. Each of the authors of this review carried out literature screening and quality evaluation independently. Our individual findings were then compared, and in the event of disagreement, a consensus was reached by means of discussion.

In this review, we identified 42 records from the literature and 54 records from other database sources that matched the appropriate keywords (Fig. 1). After eliminating duplicates, 73 records were analyzed for relevance against the topic and inclusion criteria. Twenty-four records were found to be eligible and were included

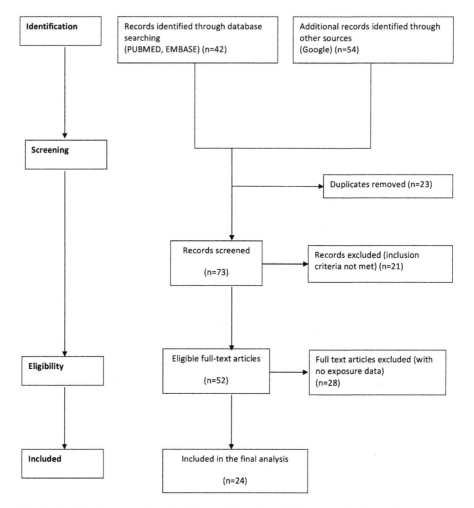

Fig. 1 Identification, screening, eligibility, and inclusion of data sources for the study

in the final analysis. Twenty-eight papers that were adjudged to be of moderate quality were excluded from the study (Fig. 1).

3   Growth and Trends in Catheterization Laboratory

The number of catheterization laboratories varies according to country, with 460 labs in India, 30 in Bangladesh, 44 in Thailand, and 50 in Malaysia (Tsapaki et al. 2011). Unfortunately, no literature on dosimetric information was available from these countries (Tsapaki et al. 2011). Approximately 3,100 operations and 725 interventional cardiac catheterizations are performed annually in the UK on babies and children who are afflicted with congenital heart disease (Petersen et al. 2003). It was documented in an earlier report that the personnel of three cardiac catheterization laboratories had performed more than 15,000 cardiac procedures over a period of 5 years in Canada (Renaud 1992). The differences in X-ray systems (old film-based systems vs. digital units) and their particular settings, levels of staff training in radiation protection, frequency of use of radiation protection facilities and personal dosimeters, and workloads of specialists all affect radiation doses received by IC staff (Vano et al. 2006a). If specialists do not regularly wear their personal dosimeters, the mean values for their occupational exposure in catheterization laboratories could provide an incorrect estimate of the real radiological risk (Vano et al. 2006; Balter 1993).

Our radiation exposure assessment shows that the badges that exceed the level 1 ALARA (as low as reasonably achievable) limits (<6 mSv/year) are indeed worn by invasive cardiologists (Andreassi et al. 2005). Even if IC staff members are involved in 1,000 angiographies per year, the annual threshold exposure level of 20 mSv is unlikely to be exceeded. An operator with a comparative 1,000 procedures per year may reach the recommended occupational limits of 150 mSv for the lens of the eye and 500 mSv for the hands (Maeder et al. 2005). In this context, we presume that IC staff is at risk in the Asian Pacific region. This is because IC staff members in the Asia Pacific region may regularly conduct 1,000 procedures annually as a result of the huge demand for these procedures in treating patients, and the high workloads of these specialists.

When laboratories possess modern radioimaging equipment, use experienced operators, and adhere to standard safety precautions, coronary intervention is considered to be quite safe for both patients and operator personnel (Efstathopoulos et al. 2003). If they are to optimally protect patients and staff, operators of X-ray instruments during catheterization procedures must know the typical dose rates for each X-ray system they use (Vano et al. 2006). The dose area product (DAP) levels for CAs and PCI in six European countries was measured as 39.1 Gycm^2 (CA) and 54.4 Gycm^2 (PCI), respectively. Based on these data, the European Research Cardiology Group for Measures for Optimising Radiological Information and Dose in Digital Imaging and Interventional Radiology has proposed temporary reference DAP levels (viz., 45 Gycm^2 for CA and 75 Gycm^2 for PCI). Several authors (Maeder et al. 2005; Kuon et al. 2003, 2004) have observed that DAP levels exceed the proposed reference levels. However, finding a correlation between the Kerama Area Product (KAP) values and the eye lens doses has been difficult (Domienik et al. 2011).

Different occupational radiation doses result from using different catheter insertion sites. The most common insertion sites for PCI utilize the femoral and radial/brachial approaches. The reason for insertion-site differential dosing during cardiac procedures is that the physician's position relative to the patient changes as the insertion site changes. The radial approach requires that the cardiologist work in closer proximity to the X-ray beam (Whitby and Martin 2005). The radial approach increased operator radiation exposure by 100% during diagnostic coronary catheterization procedures and by 50% during coronary interventions (Lange and von Boetticher 2006). No special devices or provisions were made to protect IC staff against such increased exposure. The primary reasons for the higher doses included the physician being in closer proximity to the X-ray field and longer fluoroscopy times (Kim and Miller 2009). The subclavian approach, used for implanting pacemakers and similar devices resulted in higher exposure rates than did the femoral and radial approaches, due to the operator's proximity to the X-ray beam (Limacher et al. 1988; Lindsay et al. 1992). The operator's external whole body dose was significantly higher when percutaneous transluminal coronary angioplasty (PTCA) was performed from the radial artery (13.5 ± 2.1 mrem/case), when compared to the femoral artery (8.8 ± 1.3 mrem/case) approach. By moving the floor shield to increase protection from the X-rays (3.3 ± 2.3 mrem/case vs. femoral), exposure from this procedure was reduced to levels less than that experienced from using the femoral artery approach. Thus, if proper procedures are followed, PTCA can be performed from the radial artery approach without producing increased operator radiation exposure (Mann et al. 1996).

Both physicians-in-training and staff physicians in cardiac catheterization laboratories are the groups who receive radiation doses that exceed the recommended limit. One other important factor that affects exposure level to radiation is the working attitude and techniques used by staff (Renaud 1992). In this regard, Watson et al. (1997) emphasized the importance of closely supervising cardiology fellows early in their training to limit radiation doses to patients and to staff personnel.

4   Radiation Doses for Interventional Cardiology Staff

No recommended limits for per procedure radiation doses were found in the literature. Therefore, exposure doses could not be analyzed against such threshold limits. In Fig. 2, we show the radiation dose incurred per procedure for hands/fingers/wrists by IC staff. These results show that, in ten cases, the doses received were >100 μSv. The annual radiation doses for hands/fingers/wrists among IC staff are well below the recommended dose (Fig. 3). Three observations for the radiation dose to hands exceeded the recommended ALARA 1 level (Fig. 3). The radiation dose received per procedure for the eyes and forehead of IC staff is presented in Fig. 4. Only four doses were >100 mSv. With one exception, the annual radiation exposure to eyes and forehead of IC staff were below the recommended dose (Fig 5). The recommended ALARA (level 1) of 6 mSv (Andreassi et al. 2005) for the radiation dose to eyes was exceeded in six observations (Fig. 5). Data were available for two cases that gave per-procedure doses (i.e., >100 μSv) for the thyroid/neck region of IC staff (Fig. 6). No literature was available for the annual dose received to the thyroid/neck region of IC staff.

The exceeded limits of ALARA 1 for eyes and hands were observed only among IC surgeons. The locations from which these data were gathered included a university hospital in Athens (Attikon), 34 European hospitals, and hospitals in Spain,

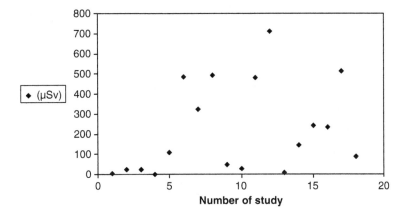

Fig. 2 Radiation dose per procedure for hands/fingers/wrists for interventional cardiology staff. Sources: Efstathopoulos et al. 2011;Tsapaki et al. 2008; Short et al. 2007; Damilakis et al. 1995; Wu et al 1991; Vano et al 1998. For each study, data points may represent combined observations from different anatomical positions (e.g., left and right wrists, fingers, and hands) and measurements with or without personal protective equipment (PPEs)

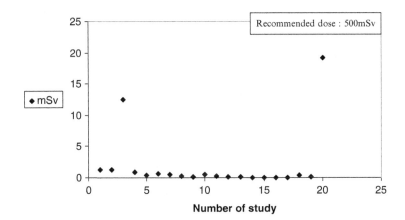

Fig. 3 Annual radiation dose for hands/fingers/wrists for interventional cardiology staff. Sources: Pyo et al. 2008; Kim et al. 2008; Efstathopoulos et al. 2011; Whitby and Martin 2005; Domienik et al. 2011. For each study, data points may represent combined observations from different anatomical positions (e.g., left and right wrists, fingers, and hands) during procedures such as Percutaneous Transluminal Coronary Angioplasty (PTCA), femoral angioplasty, stents, embolism, angiograms

Norway, Bogota (Columbia), and Montevideo (Uruguay). In the present review, we show that 12.5% (3/24) of IC surgeons experienced a risk of radiation exposure for hands and 66.7% (6/9) for eyes. This risk was based on the radiation dose observed to occur between the recommended levels and ALARA 1 level (i.e., between 6 and 500 mSv for hands and 6 and 150 mSv for eyes in different studies) (Figs. 3 and 5).

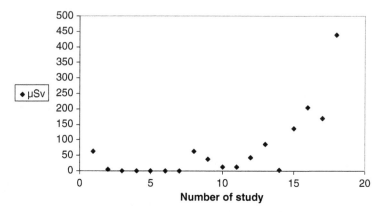

Fig. 4 Radiation dose per procedure for eyes/forehead for interventional staff. *Sources*: Efstathopoulos et al. 2011; Pratt and Shaw 1993; Calkins et al. 1991; Karppinen et al. 1995; Marshall et al. 1995; Short et al. 2007; Lie et al 2008; Wu et al 1991; Vano et al. 1998. For each study, data points may represent combined observations from different anatomical positions (e.g., left and right eyes, between the eyes) and measurements with or without PPEs

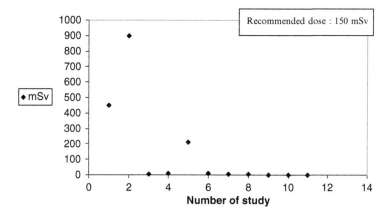

Fig. 5 Annual radiation dose for eyes/forehead for interventional cardiology staff. *Sources*: Vano et al. 1998, 2010; Pyo et al. 2008; Kim et al. 2008; Lie et al. 2008; Efstathopoulos et al. 2011; Domienik et al. 2011. For each study, data points may represent combined observations from different anatomical positions (e.g., left and right eyes, between the eyes) and measurements with or without PPEs

4.1   Hand and Wrist Exposure

Figures 2 and 3, respectively, show single procedure (μSv) and annual radiation (mSv) doses received for hands and wrists by IC staff. When IC surgeons protect their right hands by a lead screen, the radiation dose received was 147 μSv per

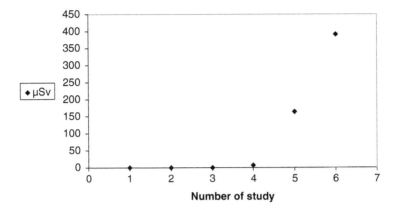

Fig. 6 Radiation dose per procedure for thyroid/neck region for interventional cardiology staff. *Sources*: Steffenino et al. 1996; Williams 1997; Calkins et al. 1991; Wu et al. 1991; Vano et al. 1998. For each study, data points may represent a combination of observations made with or without the use of PPE

procedure; without such protection the right hand received 242 μSv per procedure (Vano et al. 1998). Without a lead screen the left hand received 514 μSv, compared to 235 μSv per procedure when a lead screen was used (Vano et al. 1998). The left finger and left wrist of cardiology surgeons working in IC departments received radiation doses of 7.3 and 1.3 mSv, respectively; these doses compared to 3.6 and 1.3 mSv for the left and right wrists, respectively (Domienik et al. 2011). The right and left wrists received almost the same level of radiation per procedure in nursing/ assistant staff working in IC departments (26 μSv). However, fingers of the left hand received more radiation exposure (4 μSv) than did the fingers of the right hand (2 μSv) (Efstathopoulos et al. 2011). The annual radiation exposure level sustained by the nursing staff in IC departments was 1.3 m (Kim et al. 2008).

The left wrists of cardiology surgeons working in IC departments received a mean radiation dose of 493 μSv per procedure, whereas right wrists received a dose of 108 μSv (Efstathopoulos et al. 2011). In three earlier studies, the radiation doses to the hands of surgeons were 27, 482, and 710 μSv, respectively (Tsapaki et al. 2008; Short et al. 2007; Damilakis et al. 1995). The left fingers received more radiation than did the right fingers (324 vs. 88 μSv per procedure) (Efstathopoulos et al. 2011). Among 605 individuals performing coronary angiographies, the annual radiation dose for the left and right wrists of cardiology surgeons was 19.2 and 12.5 mSv, respectively (Efstathopoulos et al. 2011). Staff radiation doses varied between 34 and 235 μGy per procedure at the left wrist and 28 and 172 μGy at the right wrist (Goni et al. 2005). For different IC procedures performed by the same IC surgeons, the radiation dose to the hand nearer the procedure was different than that of the hand that was further away (Fig. 7; Whitby and Martin 2005).

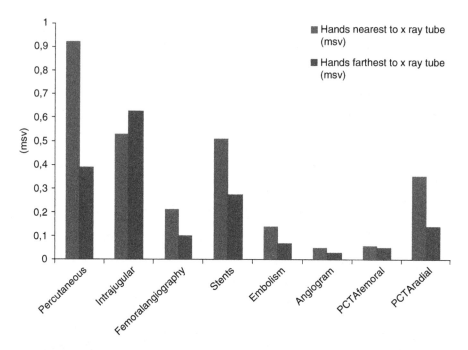

Fig. 7 Radiation dose received to the hand nearest and farthest during different IC procedures for IC surgeons (Whitby and Martin 2005)

4.2 Eye and Thyroid Gland Exposure

Figures 4 and 5 show per procedure (µSv) and annual radiation doses (mSv) for IC staff that was received to eyes and forehead, respectively. The eyes received an annual radiation dose of 1.5 mSv (Vano and Faulkner 2005), and 0.9 and 0.9 mSv (Kim et al. 2008) among IC Department nursing staff. The per-procedure radiation dose was 64 µSv (for eyes), 4 µSv (between the eyes), and 1 µSv (left eyes) (Efstathopoulos et al. 2011). A 3-year follow-up study showed the annual radiation dose received by cardiology surgeons in IC departments to be 450 mSv in 1993 and 900 mSv in 1996 (Vano et al. 1998). The per-procedure radiation dose levels for cardiology surgeons were variable: between 0.008 and 0.113 µSv (Pratt and Shaw 1993), 0.28 µSv (Calkins et al. 1991), 0.43 µSv (Karppinen et al. 1995), and 0.014 µSv (Marshall et al. 1995). The annual radiation doses recorded in various studies were as follows: 6 mSv (Vano et al. 2010), 8.2 mSv (Efstathopoulos et al. 2011), 4.1 mSv in the left eye (Domienik et al. 2011), and 3.2 mSv between the eyes (Domienik et al. 2011).

The annual radiation dose range received for 900 procedures to the eyes of cardiology surgeons was 1–11 mSv (protected eyes) and 9–210 mSv (unprotected eyes) (Lie et al. 2008). The per-procedure dose for eyes was 64 µSv between the eyes, 37 µSv in the left eye, 12 µSv between the eyes (Efstathopoulos et al. 2011), and 44 μSv for the eyes (Lie et al. 2008). For IC surgeons, the observed radiation dose for the right eye, when protected by a lead screen, was 136 μSv, whereas for the right eye, without such protection, the dose was 205 μSv; for the left eye protected by a lead screen the dose was 170 μSv, and for the unprotected left eye, the dose was 439 μSv (Vano et al. 1998). Figure 6 shows the IC staff radiation doses (μSv) received per procedure by the thyroid/neck region. The radiation dose per procedure for thyroid glands among cardiology surgeons varied across studies 0.215–0.37 μSv (Steffenino et al. 1996), 0.05–0.14 μSv (Williams 1997), 0.20 μSv (Calkins et al. 1991), 163 μSv with a lead screen (Vano et al. 1998), and 392 μSv per procedure without a lead screen (Vano et al. 1998).

5   Cataracts As an Adverse Health Effect for Interventional Cardiology Staff

A study presented at the European Society of Cardiology congress in 2009 (Duran et al. 2009) disclosed a significant difference in the frequency of lens opacity (37.9% vs. 12%, $p < 0.005$) between exposed IC staff and the control group. Therefore, the eyes may be regarded as a limiting organ for CA and PCI procedures (Lie et al. 2008). Bjelac et al. (2010) studied the prevalence of radiation-associated posterior lens opacity in Malaysia and reported an incidence of 52% (29/56, 95%CI 35–73) for interventional cardiologists, 45% (5/11, 95%CI 15–100) for nurses, and 9% (2/22, 95%CI 1–33) for control subjects. The risk of lens opacity was 5.7% (95%CI 1.5–22) for interventional cardiologists and 5.0% (95%CI 1.2–21) for nurses. Ophthalmological examinations of IC staff exposed to radiation revealed that 38% of surgeons and 21% of nurses had radiation-associated lens changes, in a study conducted in Bogota (Columbia) and Montevideo (Uruguay) from 2008 to 2009 (Vano et al. 2010). In 2004, a lens opacity of 37.3% and cataract case of 8% were observed among IC surgeons in North America (Junk et al. 2004). Yuan et al. (2010) observed that cardiologists who performed cardiac catheterization (CC) had more cataracts (1.2%) than doctors not performing CC (0.8%), in a study from all contracted hospitals of the Bureau of National Health insurance in Taiwan. However, this difference was not significant and there were several limitations to the study (Yuan et al. 2010).

If protection tools or procedures are not used, radiation doses to the lens of the eye may exceed the threshold for deterministic effects (lens opacity or cataracts) after several years of occupational exposure in this environment; this is true even when exposed staff perform as few as three to five procedures per day (Vano et al. 2008). The consequences to health of low dose occupational exposure of IC staff over long periods are still not clear. However, the role of personal protection equipment (PPE) is quite clear and is important in reducing radiation exposure (Vano et al. 1998; Lie et al. 2008).

Radiation-related cataracts tend to occur at an earlier age than senile cataracts do. Although radiation-induced cataracts may remain asymptomatic for several years, they still may impair visual function as lens opacity occurs and may produce severe and irreversible eye damage (Jacob et al. 2011). The O'CLOC (Occupational Cataracts and Lens Opacities in Interventional Cardiology) study provides further evidence about the potential risk of low-dose radiation-induced cataracts and has contributed to awareness of the importance of radiation protection among cardiologists (Jacob et al. 2011). In view of foregoing, there are concerns for the risk of radiation exposure to the lens of IC staff. Therefore, the current occupational guideline values of the International Commission on Radiation Protection (ICRP) for radiation exposure to eyes (150 my/year) may be considered as too high (Klein et al. 2009). The ICRP has reviewed recent epidemiological evidence for the lens of the eye and has issued a statement on sensitivity of eye lens tissue (ICRP 2011; Rehani et al. 2011). For occupational exposures that are planned, the commission now recommends an equivalent annual dose limit for the lens of 20 mSv, averaged over defined periods of 5 years, with no single annual exposure exceeding 50 mSv (ICRP 2011; Rehani et al. 2011).

Interventional radiologists need 20/20 vision in both eyes to maintain excellent stereopsis and to perform the delicate procedures demanded by their job. Treatment and surgery for cataracts is a frequent and successful surgical procedure. But, risks are associated with cataract surgery that can negatively affect outcomes and may affect visual rehabilitation prospects for interventional radiologists (Haskal 2004). The use of a mechanical injector pump for coronary arteriography has reduced the radiation exposure of cardiologists (Grant et al. 1993). This technique is safe, convenient, produces angiograms of comparable quality to hand injection and should be recommended as standard practice to reduce radiation (Grant et al. 1993). Earlier studies have confirmed that by using this pump, radiation exposure during cervical irradiation was reduced by a factor of 15; in addition, radiation exposure to the left wrist was reduced by a factor of 8 by using the protection afforded by a suspended lead screen during irradiation (Wyart et al. 1997).

The sensitivity of lens tissue to radiation damage makes eye protection essential for medical personnel (Pratt and Shaw 1993). Cardiologists often fail to routinely use protective leaded eyewear; it raises the crucial need of staff to wear radiation monitoring devices to prevent cataracts and protect the eyes (Vano and Faulkner 2005). It was observed that under high workload conditions, where inadequate protective measures were used, it is possible for the operator eye dose to exceed the recommended level (set at 3/10 of the 150 mSv dose limit) (Jeans et al. 1985). Reducing radiation-related cataract risks among interventional medical personnel can be achieved by the effective use of protective devices (Rehani et al. 2011). A systematic radiological protection training program (installing a radiation badge policy for staff) can improve compliance by 36–77% (McCormick et al. 2002). A strict policy to regularly use personal dosimeters should be part of any safety or quality program in cardiology laboratories (McCormick et al. 2002).

The relationship between the radiation eye dose received by cardiologist's, and factors such as the dose efficiency of the X-ray equipment, effects of scattered dose rates, examination protocols, and workload are complex and may vary from center to center (Pratt and Shaw 1993). The procedure or techniques used, the catheter or catheter insertion site chosen, operator positioning and the appropriate use of personal protective devices also play an important role in the levels of radiation exposure sustained (Kim and Miller 2009). The dose rate at eye level decreases by a factor of 2 when the physician's eye level above the floor increases from 1.6 to 1.8 m (Pratt and Shaw 1993). When a physician stands at the patient's groin on the right-hand side and uses a femoral approach, the left anterior oblique (LAO) cranial projection results in the highest operator dose rate for scattered radiation (Pratt and Shaw 1993; Kuon et al. 2004; Camm et al. 1993). Awareness among cardiologists about the radiation risk of the LAO projection may produce a radiation dose that is two to three times lower (Pitney et al. 1994).

6   Other Adverse Health Effects

Concerns exist about the low-dose radiation (LDR) health effects among IC staff. Chronic exposure to the effects of LDR is known to increase hydrogen peroxide levels in IC staff and to alter redox balance (Russo et al. 2011). Russo et al. (2011) described two adaptive cellular responses to the effects of irradiation: (1) enhanced antioxidant defense (increases in glutathione, counteracting increased oxyradical stress) and (2) increased susceptibility to apoptotic induction, which might efficiently remove genetically damaged cells. Venneri et al. (2009) suggested that cumulative professional radiological exposure is associated with a non-negligible lifetime attributable risk of cancer for those cardiac catheterization laboratory staff that have the most exposure (Venneri et al. 2009). Such exposure was also associated with an increased micronuclei frequency for interventional cardiologists, but not for clinical cardiologists, which risk correlates with years of professional activity (Andreassi et al. 2005).

Interventional cardiologists, who have an average exposure of 4 mSv/year, show a twofold increase in certain biomarkers like circulating lymphocytes, chromosome aberrations, and/or micronuclei. The appearance of these biomarkers in IC staff are surrogates of cancer risk and could represent intermediate carcinogenesis end points (Andreassi et al. 2005; Zakeri and Assaei 2004; Maffei et al. 2004). Hence, interventional cardiology is recognized as being a practice that has high radiation risk (Finkelstein 1998; Kim et al. 2008; Vano 2003). We again underscore that monitoring for levels of occupational exposure should become an important part of any and all quality assurance (QA) programs that are established for such practices (Vano et al. 2006).

7   Need for Radiation Safety Practices in Catheterization Laboratories

As interventional procedures have increased in numbers, radiation exposure to personnel working in cardiac catheterization laboratories has increased. The current radiation precautions appear to be adequate, because the radiation dose reported for

IC staff is low. However, in view of the rather high incidence of cataracts reported among IC staff, there is a need for strict implementation of radiation safety practices in the medical workplace. Programs should be implemented to initiate safety awareness and to provide for radiation protection training among IC staff (Vano 2003; Tsapaki et al. 2011). Another important undertaking is to institute the routine use of dosimeters among staff during IC procedures and to regularly perform surveillance of occupational doses for the whole body and for eyes, thyroid glands, and hands (Whitby and Martin 2005). Only instruments that comply with radiation safety standards and practices should be used during IC procedures, so that radiation exposures are optimized (Yuan et al. 2010). Of most importance in radiation safety is the regular use of personal protective equipment or shielding for staff in the workplace (Tsapaki et al. 2011; Kim et al. 2008; Pratt and Shaw 1993). Working at a safe distance from instruments, and assuring that such instruments are properly positioned can reduce the radiation dose received by IC staff (Maeder et al. 2005; Vano et al.1998; Jeans et al. 1985).

This review further underscore the importance of each catheterization laboratory undertaking routine measurement of the dose received by each IC staff member. Our review revealed that, in most of the available literature, data were given as radiation dose per procedure, and, unfortunately, no recommended dose rate has been established for individual procedures. Depending on the type of procedure and the technique used, the operator dose per procedure may range from 3 to 450 μSv at the neck, from less than 0.1 to 32 μSv at the waist or chest, and from 48 to 1,280 μSv at the hands (Miller et al. 2010). Translating such exposure values into monthly or annual worker dose limits is difficult (Miller et al. 2010). More and better monitoring of radiation doses among catheterization laboratory IC staff is needed for hands, eyes, and thyroid glands. This can be accomplished by using personal dosimeters and by developing recommended limits for IC staff per-procedure doses. Finally, we suggest that the national and international agencies (e.g., ICRP) that are responsible for medical and radiation safety do work to establish such radiation exposure limits for all relevant IC protocols.

8 Summary

To the best of our knowledge, this chapter constitutes the first systematic review of radiation exposure to eyes, thyroid, and hands for Interventional Cardiology (IC) staff. We have concluded from our review that these anatomical locations are likely to be exposed to radiation as a result of the limited use of personal protective equipment (PPE) among IC staff as shown in Fig. 8. Our review also reveals that, with the exception of three eye exposure cases, the annual radiation dose to eyes, thyroid, and hands among IC staff was within recommended levels and limits. The As Low As Reasonably Achievable (ALARA) limit was not achieved in three cases for fingers/hands and four cases for eyes. However, an increased incidence of cataracts were reported for IC staff, and this gives rise to the concern that low-dose or

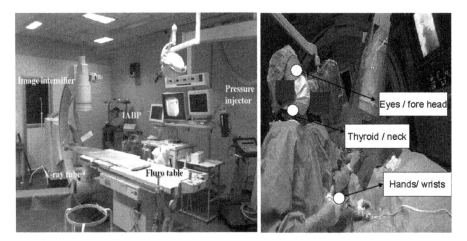

Fig. 8 Cardiac Catheterization laboratory or cath labs and the possible radiation exposure routes for IC staff. *IABP* Intraaortic balloon pump. *Source*: Cardiac catheterization laboratory 2011

unnoticed exposures may increase the risk of developing cataracts among cardiology staff. Clearly, the formation of cataracts among IC staff may be an issue and should be studied in more depth.

Our review also disclosed that the two groups who receive excessive radiation doses (i.e., exceed the recommended limit) are physicians-in-training and junior staff physicians who work in cardiac catheterization laboratories. In particular, more attention should be given to assessing the effects of radiation exposure among IC staff who work in the Asia Pacific countries, because our review indicates that the number of IC procedures performed by IC staff in these countries is higher than for other continents. There is a huge demand for procedures conducted by IC staff in the Asia-Pacific area, for both treating patients and consulting with specialists.

Our review also disclosed that recommended limits for per-procedure radiation doses are needed for IC staff. We recommend that such limits be established by the appropriate national and international agencies that are responsible for occupational radiation exposure. Although our review indicates that the current precautions against LDR exposure for IC staff are adequate in most cases, we are concerned about the relatively high incidence of cataracts reported to exist among IC staff. Therefore, we believe that there is a need for a strict implementation of radiation safety practices in cardiology laboratories and associated workplaces that utilize radiation.

The action that is most important for protecting staff in the workplace against radiation exposure is the regular use of personal protective equipment or shielding. Working at a safe distance from instruments and assuring that such instruments are in the proper position are other techniques that can reduce the radiation dose received by IC staff.

Acknowledgments We wish to thank Dana Wendeler, Documentation officer, Department of Occupational Health Research, Institute for Statutory Accident Insurance and Prevention in the Health and Welfare Services (BGW), Hamburg, Germany, for her support with the management of the literature. This research project was funded by the Institute for Statutory Accident Insurance and Prevention in the Health and Welfare Services (BGW), Hamburg, Germany.

References

Andreassi MG, Cioppa A, Botto N, Joksic G, Manfredi S, Federici C, Ostojic M, Rubino P, Picano E (2005) Somatic DNA damage in interventional cardiologists: a case–control study. FASEB J19(8):998–9

Asian Network of Cardiologists in Radiation Protection—under RCA/IAEA project (2007) Newsletter (Issue N1): 1–2. Available via http: 77rpop.iaea.org/RPOP/RPoP/Content/ AdditionalResources/Training/2_TrainingEvents/asian-network.htm. Accessed 23 Nov 2011

Bjelac CO, Rehani MM, Sim KH, Liew HB, Vano E, Kleiman NJ (2010) Risk for radiation induced cataract for staff in interventional cardiology: Is there reason for concern? Catheter Cardiovasc Interv 76:826–834

Balter S (1993) Guidelines for personnel radiation monitoring in the cardiac catheterisation laboratory. Cathet Cardiovasc Diagn 30:277–279

Calkins H, Niklason L, Sousa J, El-Atassi R, Langberg J, Morady F (1991) Radiation exposure during radiofrequency catheter ablation of accessory atrio ventricular connections. Circulation 84:2376–82

Camm AJ, Reid J, Raphael M, Wilde P, Boyle R, Clarke M, Qureshi S, Rothman M, Shaw A (1993) Radiation hazards to the cardiologist—a report of a Subcommittee of the British Cardiac Society. Br Heart J 70:489–496

Cardiac catheterisation laboratory (2011) Available via http://cardiophile.org/common/2011/11/ cardiac-catheterisation-laboratory-place-where-angiography-and-interventions-are-done. Accessed 12 Dec 2011

Dawkins KD, Gershlick T, de Belder M (2005) Percutaneous coronary intervention: Recommendations for good practice and training. Heart 91(6):1–27

Damilakis J, Koulourakis M, Hatjidakis S, Karabekios S, Gourtsoyiannis N (1995) Radiation exposure to the hands of operators during angiographic procedures. Br J Radiol 21:72–5

Domienik J, Brodecki M, Carinou E, Donadille L, Jankowski J, Koukorava C, Krim S, Nikodemova D, Ruiz Lopez N, Sans Merce M, Struelens L, Vanhavere F (2011) Extremity and eye lens doses in interventional radiology and cardiology procedures. First results of the oramed project. Radiat Prot Dosim 144(1–4):442–447

Delichas M, Psarrakos K, Molyvda-Athanassopoulou E, Giannoglou G, Sioundas A, Hatziioannou K, Papanastassiou E (2003) Radiation exposure to cardiologists performing interventional cardiology procedures. Eur J Radiol 48:268–273

Duran D, Duran G, Ramirez R, Vano E, Kleinman N, Echeverri D, Gomez G, Cabrera M (2009) Cataracts in interventional cardiology personnel. Retrospective evaluation study of lens injuries and dose (RELID Study). Eur Heart J 30:872

Efstathopoulos E, Makrygiannis SS, Kottou S, Karvouni E, Giazitzoglou E, Korovesis S, Tzanalaridon E, Rapton PD, Katritsis DG (2003) Medical personnel and patient dosimetry during coronary angiography and intervention. Phys Med Biol 48:3059–68

Efstathopoulos EP, Pantos I, Andreou M, Gkatzis A, Carinou E, Koukorava C, Kelekis NL, Brountzos E (2011) Occupational radiation doses to the extremities and the eyes in interventional radiology and cardiology procedures. Br J Radiol 84:70–77

Finkelstein MM (1998) Is brain cancer an occupational disease of cardiologists? Can J Cardiol 14:1385–1388

German Guidelines (2003) Bekanntmachung der Neufassung der Röntgenverordnung vom 30. April 2003. Bundesgesetzblatt 2003; Teil 1:604–635

Goni H, Papadopoulou D, Yakoumakis E, Stratigis N, Benos J, Siriopoulou V, Makri T, Georgiou E (2005) Investigation of occupational radiation exposure during interventional cardiac catheterisations performed via radial artery. Radiat Prot Dosim 117:107–110

Grant SCD, Faragher EB, Hufton AP, Bennett DH (1993) Use of a remotely controlled mechanical pump for coronary arteriography—a study of radiation exposure and quality implications. Br Heart J70:479–484

Hafez MA, Smith RM, Matthews SJ, Kalap G, Sherman KP (2005) Radiation exposure to the hands of orthopaedic surgeons: are we underestimating the risk? Arch Orthop Trauma Surg 125:330–335

Haskal ZJ (2004) Interventional radiology carries occupational risk for cataracts. RSNA News 14:5–6

International Commission on Radiological Protection (2000) Avoidance of radiation injuries from medical interventional procedures. ICRP Publication 85. Ann ICRP 230:7–67

International Commission on Radiological Protection (2011) Tissue reactions and other non-cancer effects of radiation. Elsevier. Ann ICRP Ref 4834-1783-0153

Junk A, Haskal Z, Worgul B (2004) Cataract in interventional radiology—An occupational hazard? Invest Ophthalmol Vis Sci 388:45

Jacob S, Bertrand A, Bernier MO (2011) Occupational cataracts and lens opacities in interventional cardiology: The O'CLOC study. Towards convergence of technical nuclear safety practice in Europe. Available via http://www.eurosafe-forum.org/userfiles/2_10_Paper_EUROSAFE_SJ.pdf. Accessed 4 Dec 2011

Jeans SP, Faulkner K, Love HG, Bardsley RA (1985) An investigation of the radiation-dose to staff during cardiac radiological studies. Br J Radiol 58:419–428

Kim KP, Miller DL (2009) Minimising radiation exposure to physicians performing fluoroscopically guided cardiac catheterisation procedures: A review. Rad Prot Dosimetry 133:227–233

Kim KP, Miller DL, Balter S, Kleinerman RA, Linet MS, Kwon D (2008) Occupational radiation doses to operators performing cardiac catheterisation procedures. Health Phys 94:211–227

Karppinen J, Parviainen T, Servomaa A, Komppa T (1995) Radiation risk and exposure of radiologists and patients during coronary angiography and PTCA. Radiat Prot Dosim 57:481–5

Klein LW, Miller DL, Blater S, Laskey W, Haines D, Norbash A, Mauro MA, Goldstein JA, Joint Inter-Society Task Force on Occupational Hazards in the Interventional Laboratory (2009) Occupational health hazards in the interventional laboratory: Time for a safe environment. J Vasc Interv Radiol 20(7 Suppl):S278–83

Kottou S, Neofotistou V, Tsapaki V, Lobotessi H, Manetou A, Molfetas MG (2001) Personnel doses in haemodynamic units in Greece. Radiat Prot Dosim 94:121–124

Kuon E, Günther M, Gefeller O, Dahm JB (2003) Standardization of occupational dose to patient DAP enables reliable assessment of radiation-protection devices in invasive cardiology. Rofo 175:1545–50

Kuon E, Empen K, Rohde D, Dahm JB (2004) Radiation exposure to patients undergoing percutaneous coronary interventions. Are the current reference values too high? Herz 29:208–17

Lie OO, Paulson GU, Wohni T (2008) Assessment of the effective dose and dose to the lens of the eye for the interventional cardiologist. Radiat prot Dosimetry 132(3):313–318

Lange HW, von Boetticher H (2006) Randomized comparison of operator radiation exposure during coronary angiography and intervention by radial or femoral approach. Catheter Cardiovasc Interv 67:12–16

Limacher MC, Douglas PS, Germano G, Laskey WK, Lindsay BD, McKetty MH, Moore ME, Park JK, Prigent FM, Walsh MN (1988) Radiation safety in the practice of cardiology. J Am Coll Cardiol 31:892–913

Lindsay BD, Eichling JO, Ambos HD, Cain ME (1992) Radiation exposure to patients and medical personnel during radiofrequency catheter ablation for supraventricular tachycardia. Am J Cardiol 70:218–223

Maffei F, Angelini S, Forti GC, Violante FS, Lodi V, Mattioli S, Hrelia P (2004) Spectrum of chromosomal aberrations in peripheral lymphocytes of hospital workers occupationally exposed to low doses of ionizing radiation. Mutat Res 547:91–9

Maeder M, Verdun FR, Stauffer JC, Ammann P, Rickli H (2005) Radiation exposure and radiation protection in interventional cardiology. Kardiovaskuläre Medizin 8:124–132

Moher D, Liberati A, Tetzlaff J, Altman DG (2009) Preferred reporting items for systematic reviews and meta-analyses: the PRISMA statement. PLoS Med6:e1000097. doi:10.1371/journal. pmed.1000097

Marshall NW, Noble J, Faulkner K (1995) Patient and staff dosimetry in neuroradiological procedures. Br J Radiol 68:495–501

Miller DL, Vano E, Bartal G, Balter S, Dixon R, Padovani R, Schueler B, Cardella JF, de Baere T (2010) Occupational radiation protection in interventional radiology: A joint guideline of the Cardiovascular and Interventional Radiology Society of Europe and the Society of Interventional Radiology. Cardiovasc Intervent Radiol 33:230–239

Mann JT, Cubeddu G, Arrowood M (1996) Operator radiation exposure in PTCA: comparison of radial and femoral approaches. J Invasive Cardiol 8:D22–D25

McCormick VA, Schultz CC, Hollingsworth-Schuler V, Campbell JM, O'Neill WW, Ramos R (2002) Reducing radiation dose in the cardiac catheterization laboratory by design alterations and staff education. Am J Cardiol 90:903–905

Picano E, Santoro G, Vano E (2007) Sustainability in the cardiac cath lab. Int J Cardiovasc Imaging 23(2):143–7

Pratt TA, Shaw AJ (1993) Factors affecting the radiation dose to the lens of the eye during cardiac catheterisation procedures. Br J Radiol 66:346–350

Petersen S, Peto V, Rayner M (2003) Congenitial heart disease statistics 2003. British Heart Foundation statistics. Available via http://www.bhf.org.uk/idoc.ashx?docid=b41e291d-0dec-41f6-b726-10b609299aa&version=−1. Accessed 13 Aug 2011

Pitney MR, Allan RM, Giles RW, Mclean D, Mccredie M, Randell T, Walsh WF (1994) Modifying fluoroscopic views reduces operator radiation exposure during coronary angioplasty. J Am Coll Cardiol 24:1660–1663

Raza SMS (2011) Radiation exposure in the cath lab-safety and precautions. Available via http://priory.com/med/radiation.htm. Accessed 20 July 2011

Rotter M, Pfiffner D, Maier W, Zeiher AM, Meier B (2003) Working Group Interventional Cardiology and Coronary Pathophysiology, European Society of Cardiology. Interventional cardiology in Europe 1999. Eur Heart J24:1164–70

Renaud L (1992) A 5-year follow-up of the radiation exposure to in-room personnel during cardiac catheterization. Health Phys 62:10–15

Rehani MM, Vano E, Bjelac OC, Kleiman NJ (2011) Radiation and cataract. Radiat Prot Dosim. doi: 10.1093/rpd/ncr299

Russo LG, Tedesco I, Russo M, Cioppa A, Andreassi MG, Picano E (2011) Cellular adaptive response to chronic radiation exposure in interventional cardiologists. Eur Heart J23. doi: 10.1093/eurheart/ehr263

Sim KH, Rehani M, Kleiman N, Bjelac OC, Vano E (2010) Radiation induced lens opacities in the eyes of cath lab staff. J Am Coll Cardiol 55(A201):E1888

Short CP, Al Hashinii H, Malone L, Lee MJ (2007) Staff radiation doses to the lower extremities in interventional radiology. Cardiovasc Intervent Radiol 30:1206–9

Steffenino G, Rossetti V, Dellavalle A, Garbarino M, Cerati R, Norbiato A, Uslenghi E (1996) Staff dose reduction during coronary angiography using low framing speed. Br J Radiol 69:860–4

Tsapaki V, Ghulam MF, Lim ST, Minh HN, Nwe N, Sharma A, Sim KH, Srimahachota S, Rehani MM (2011) Status of radiation protection in various interventional cardiology procedures in the Asia Pacific region. Heart Asia 3:16–24

Tsapaki V, Patsilinakos S, Voudris V, Magginas A, Pavlidis S, Maunis T (2008) Level of patient and operator dose in the largest cardiac centre in Greece. Radiat Prot Dosimetry 129:71–3

Vano E, Gonzalez L, Beneytez F, Moreno F (1998a) Lens injuries induced by occupational exposure in non-optimized interventional radiology laboratories. Br J Radiol 71:728–33

Vano E, KleimanNJ DA, Rehani MM, Eche D, Cabrera M (2010a) Radiation cataract risk in interventional cardiology personnel. Rad Res 174(4):490–495

Vano E, Gonzalez L, Guibelalde E, Fernandez JM, Ten JI (1998b) Radiation exposure to medical staff in interventional and cardiac radiology. Br J Radiol 71:954–60

Vano E, Kleiman NJ, Duran A, Rehani MM, Echeverri D, Cabrera M (2010b) Radiation cataract risk in interventional cardiology personnel. Rad Res 174(4):490–495

Vano E (2003) Radiation exposure to cardiologists: how it could be reduced. Heart 89:1123–4

Vano E, Gonzalez L, Fernandez JM (2008) Eye lens exposure to radiation in interventional suites: caution is warranted. Radiology 248(3):945–953

Vano E, Gonzalez L, Fernandez JM, Alfonso F, Macaya C (2006a) Occupational radiation doses in interventional cardiology: a 15 year follow up. Br J Radiol 79:383–388

Vano E, Gonzalez L, Fernandez JM, Prieto C, Guibelade E (2006b) Influence of patient thickness and operation modes on occupational and patient radiation doses in interventional cardiology. Radiat Prot Dosimetry 118(3):325–330

Vano E, Faulkner K (2005) ICRP special radiation protection issues in interventional radiology, digital and cardiac imaging. Radiat Protect Dosim 117(1–3):13–17

Venneri L, Rossi F, Botto N, Andreassi MG, Salcone N, Emad A, Lazzeri M, Gori C, Vano E, Picano E (2009) Cancer risk from professional exposure in staff working in cardiac catheterisation laboratory: Insights from the National Research Council Biological effects of ionizing radiation VII report. Am Heart J157:118–24

Watson LE, Riggs MW, Bourland PD (1997) Radiation exposure during cardiology fellowship training. Health Phys 73:690–693

Wu JR, Huang TY, Wu DK, Hsu PC, Weng PS (1991) Radiation exposure of pediatric patients and physicians during cardiac catheterisation and balloon pulmonary valvuloplasty. Am J Cardiol 68:221–225

Whitby M, Martin CJ (2005) A study of the distribution of dose across the hands of interventional radiologists and cardiologists. Br J Radiol 78:219–229

Williams JR (1997) The interdependence of staff and patient doses in interventional radiology. Br J Radiol 70:498–503

Wyart P, Dumant D, Gourdier M, Nassar F, Bouthillon JC, Chestier Y (1997) Contribution of self-surveillance of the personnel by electronic radiation dosimeters in invasive cardiology. Arch Mal Coeur Vaiss 90:233–238

Yuan MK, Chien CW, Lee SK, Hsu NW, Chang SC, Chang SC, Tang GJ (2010) Health effects of medical radiation on cardiologists who perform cardiac catheterisation. J Chin Med Assoc 73(4):199–204

Zakeri F, Assaei RG (2004) Cytogenetic monitoring of personnel working in angiocardiography laboratories in Iran hospitals. Mutat Res 562:1–9

Environmental Fate and Toxicology of Methomyl

April R. Van Scoy, Monica Yue, Xin Deng, and Ronald S. Tjeerdema

Contents

A.R. Van Scoy (✉) • R.S. Tjeerdema
Department of Environmental Toxicology, College of Agricultural and Environmental Sciences,
University of California, One Shields Ave, Davis, CA 95616-8588, USA
e-mail: avanscoy@ucdavis.edu

M. Yue • X. Deng
Department of Pesticide Regulation, California Environmental Protection Agency,
Sacramento, CA 95812-4015, USA

D.M. Whitacre (ed.), *Reviews of Environmental Contamination and Toxicology,*   93
Reviews of Environmental Contamination and Toxicology 222,
DOI 10.1007/978-1-4614-4717-7_3, © Springer Science+Business Media New York 2013

1 Introduction

The insecticide methomyl (*S*-methyl *N*-(methylcarbamoyloxy)thioacetimidate; CAS 16752-77-5; Fig. 1) was first introduced by E.I. du Pont de Nemours in 1968 (US EPA, 1998b). In 1978, the US Environmental Protection Agency classified methomyl as a restricted-use pesticide (RUP; US EPA 1998a); currently 15 registered products are categorized as such (US EPA 1998b). Further restrictions were implemented in 1995, limiting use to certain agricultural production areas, requiring addition of an embittering agent during formulation and requiring the use of bait stations (US EPA 1998a). Within the USA, approx. 262,000 kg of methomyl (a.i.) was applied on agricultural crops annually from 1999 to 2004 (US EPA 2010). However, estimates for the period between 2001 and 2007 show annual average usage of approx. 363,000 kg (a.i.); major crop uses included sweet corn, lettuce, onions, and tomatoes (US EPA 2010). In 2007, some 227,711 kg of active ingredient was applied in California alone (CDPR 2007).

Methomyl is an oxime carbamate insecticide that controls a broad spectrum of arthropods such as spiders, ticks, moths, flies, beetles, aphids, leafhoppers, and spider mites often found on various field crops, ranging from fruits to tobacco (Kidd and James 1991). Methomyl is formulated as a soluble concentrate, a wettable powder or a water-soluble powder (Kidd and James 1991) and is the active ingredient of Du Pont 1179™, Flytek™, and Kipsin™, among other trade formulations (Kamrin and Montgomery 1999). Furthermore, the main formulated water-soluble products contain approx. 25–90% methomyl, whereas the water-miscible products only contain some 12.5–29% (IPCS 1995). Methomyl is weak-to-moderately persistent, with a soil half life ($t_{1/2}$) ranging from a few to more than 50 days; however, under ideal field conditions the $t_{1/2}$ should be no longer than 1 week (IPCS 1995).

Human exposures to methomyl fall into three toxicity categories defined by the US EPA that depend on the route of exposure: I, oral exposure (highly toxic); II, inhalation (moderately toxic); and III, dermal exposure (slightly toxic; US EPA 1998b). Furthermore, methomyl is considered to be highly toxic to mammals, fish and aquatic invertebrates (Farre et al. 2002). To illustrate, the acute oral LC_{50} given for rats was 17–45 mg/kg (Mahgoub and El-Medany 2001), the LC_{50} values for bluegill sunfish and rainbow trout were 0.9–3.4 mg/L, and the LC_{50} values for *Daphnia magna* were from 0.022 to 0.026 mg/L (Yi et al. 2006; Periera et al. 2009). Because methomyl's water solubility and toxicity to non-targeted aquatic organisms is high (Table 1), concerns exist for its potential impact on surface water, groundwater, and aquatic organisms. Therefore, the most up to date information may be useful

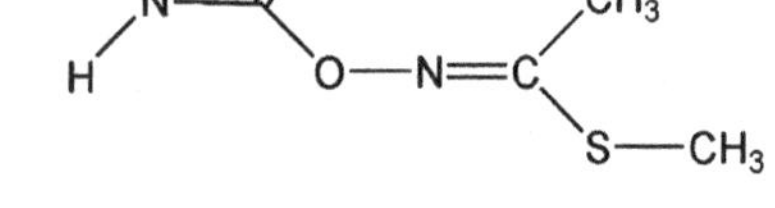

Fig. 1 Methomyl structure

Table 1 Physicochemical properties of methomyl

Chemical Abstract Service registry number (CAS#)[a]	16752-77-5
Molecular formula[a]	$C_5H_{10}N_2O_2S$
Molecular weight (g/mol)[a]	162.2
Density at 25°C (g/mL)[a]	1.29
Melting point (°C)[a]	78–79
Octanol-water partition coefficient (log K_{ow})[b]	1.24
Organic carbon normalized partition coefficient (K_{oc})[b]	72
Vapor pressure at 25°C (mmHg)[b]	5.6×10^{-6}
Henry's law constant (Pa m³ mol⁻¹)[a]	2.13×10^{-6}
Solubility at 25°C (g/L)[a]	
Water	57.9
Methanol	1,000
Acetone	730
Ethanol	420
Isopropanol	220
Toluene	30

[a]Data from Tomlin (2000), [b]Data from US EPA (1989)

in characterizing any potential environmental effects attributable to methomyl. To that end, we have reviewed the relevant literature, and in this chapter address methomyl's chemistry, environmental fate, and toxicology.

2 Chemistry

Methomyl is an *O*-(methylcarbamoyl)oxime carbamate; as such, its structure is similar to both aldicarb and thiocarboxime (Kuhr and Dorough 1976). When pure, methomyl is a white crystalline solid with a slight sulfurous odor. At room temperature, it is moderately to highly soluble in water and alcohols and has a low affinity for both soils (e.g., illite) and organic matter. Methomyl is denser than water, is susceptible to hydrolysis under alkaline conditions, and is subject to degradation via photocatalytic reactions and by microbes at various rates. Methomyl's physicochemical properties are presented in Table 1.

3 Chemodynamics

3.1 Soil

Because of its strongly hydrophilic nature, there is concern that methomyl may contaminate both surface and groundwater. Although increased soil organic matter and clay content (both amount and type) can influence methomyl's retention by soil, its overall adsorption to soil is generally weak-to-moderate at best.

Several researchers have assessed the adsorption of methomyl by various soil types and organic compositions. For example, Cox et al. (1993) investigated the role that clays (smectites, illites, and kaolinites) and humic acids (saturated with cations) play in methomyl sorption. In general, perhaps because of its surface area, the sorption to smectites (K_d=4.5–9.58) was greater than to both illites (K_d=1.56) and kaolinites (K_d=0.5). Methomyl was shown to also possess a higher sorption affinity for humic acid (K_d=399.5) than clays (Cox et al. 1993).

Leistra et al. (1984) calculated sorption coefficients to model the extent of methomyl leaching in greenhouse soils (sandy, loamy sand, and loam soil), under different transformation (degradation) and irrigation rates. They found only 0.03% of the original mass had leached after 110 days, under both low transformation (first-order rate coefficient k_r=0.0495 day^{-1}; $t_{1/2}$ = 14 day) and high irrigation (4 mm/day) rates; thus minimal leaching of the insecticide was predicted from these results. Furthermore, adsorption coefficients for soil/liquid partitioning (K_d) were determined. The resulting coefficients, 0.46×10^{-3} m^3/kg (sandy), 0.43×10^{-3} m^3/kg (loamy sand), and 1.30×10^{-3} m^3/kg (loam) indicate that methomyl has a weak-to-moderate affinity for soils (Leistra et al. 1984).

Jones et al. (1989) reported methomyl to have a $t_{1/2}$ of 2 days in surface soils and 0.5–1.6 months in subsoils. However, values reported in other studies were different; under laboratory conditions Kahl et al. (2007) reported an average $t_{1/2}$ of 15.5 days in topsoil, whereas under field conditions the $t_{1/2}$ was approx. 0.97–1.25 days for cropped soil (Aktar et al. 2008). To summarize, although predictions vary with soil type and organic matter content, they all indicate that methomyl is not very persistent in complex soils.

Variations in reported adsorption coefficients and half-lives indicate that environmental conditions are important in influencing this pesticide's transport (i.e., leaching) and degradation. Because methomyl has been widely used in agriculture, it is important to understand its transport and fate within field soils. It is known to be rapidly degraded into CO_2 by soil microbes (Nyakundi et al. 2011); however, trace amounts of the parent insecticide and its hydrolytic product (S-methyl-N-hydroxythioacetamidate) are also detectable (Harvey and Pease 1973). Furthermore, Nyakundi et al. (2011) demonstrated the potential of white rot fungi to remediate the insecticide in contaminated soils. Kahl et al. (2007) investigated the depth to which methomyl can leach in soil. The highest concentrations appeared at an 80 cm depth, with degree of transport dependent on water flow and degree of soil porosity.

3.2   Water

Methomyl has high water solubility and a weak-to-moderate adsorption to soils, and therefore poses a contamination risk to surface and groundwater (Table 1). The US

Geological Survey's National Water-Quality Assessment (NAWQA) Program monitored eight US urban surface waters for residues of herbicides and insecticides (Hoffman et al. 2000), methomyl residues were detected only in Las Vegas Wash (Las Vegas, NV); the probable source of these detections was sewage treatment plant effluent and urban runoff (Hoffman et al. 2000). NAWQA also analyzed for pesticide residues in groundwater between 1992 and 1996. They sampled 2,485 sites and detected residues of 67 pesticides. The maximum methomyl concentration detected in this study was <17 ng/L (Kolpin et al. 2000). In California (CDPR 2011), measurable methomyl concentrations were found in many monitored urban and agricultural waterways. The highest residue detected (55.3 μg/L) was from Chualar Creek (Monterey County), whereas the highest agricultural-related detection (0.343 μg/L) was in Orcutt Creek (Santa Barbara County; CDPR 2011). The risks posed in these and other locations can be established from aquatic life benchmarks that have been set by the US EPA; invertebrates are considered the most sensitive species and have the lowest chronic aquatic life benchmark. The residue levels found in Orcutt Creek were below the chronic aquatic invertebrate benchmark (0.7 μg/L), which suggests a low exposure risk. In contrast, exposure risks in Chualar Creek were relatively high, thus increasing the potential for nontarget species toxicity.

The leaching of methomyl and its degradation products into water sources has also been investigated. To evaluate the effects of both irrigation and rainfall, the insecticide thiodicarb was applied to two sites containing sandy clay loam and sandy loam soils, respectively; each site was regarded to posses the potential for ground-water contamination (Jones et al. 1989). Since methomyl is the principal degradation product of thiodicarb (which is also hydrophobic), there is a presumption that it will be detected at higher levels within water sources. Jones et al. (1989) found that soil collected 2 months after application contained low residue levels; however, methomyl was detected 1 month after application in groundwater at depths of 1.8 and 3.2 m. Carbo et al. (2008) also examined the potential of methomyl to contaminate shallow aquifers (<4.5 m deep) in Mato Grosso, Brazil. Water samples collected from monitoring wells placed in cotton fields contained measurable concentrations, ranging from less than the limit of detection (LOD=0.10 μg/L) to 22.81 μg/L (Carbo et al., 2008).

During any rainfall event, methomyl has the potential to run off of application sites and into adjacent uncontaminated sites. To investigate this phenomenon, Harvey and Pease (1973) studied a loamy sand soil that had been cleared of vegetation. The levels of insecticide detected in the runoff (<0.01 mg/L) and in the soil from both treated and untreated plots (<0.04 mg/L, 15 days after application), suggested that little surface runoff or leaching occurs providing the agent is applied at recommended rates. Furthermore, Kahl et al. (2008) detected dissolved concentrations (i.e., 11.4% of applied mass; water-pesticide suspension applied via spraying) in stream water that were four times greater than that of the fungicide chlorothalonil (which is strongly hydrophobic; detected at levels of 3.5% of applied mass in the same stream water).

3.3   Air

The volatilization rate of methomyl from both dry surfaces and water is relatively low, as predicted by its low vapor pressure and Henry's law constant values (Table 1). Hence, volatilization does not significantly contribute to methomyl's dissipation from soil (either moist or dry) or water. Yeboah and Kilgore (1984) determined that the methomyl concentrations (from liquid concentrate) measured in the ambient air of a pesticide storage building were minimal (13.7 ng/m^3), compared to its threshold limit value (TLV = 2,500 µg/m^3; ACGIH 1982; Yeboah and Kilgore 1984). When compared to other monitored pesticides, Baker et al. (1996) reported that concentrations in ambient air near methomyl-treated crops were non-detectable.

4   Environmental Degradation

4.1   Abiotic Processes

Hydrolysis
Methomyl is potentially subject to hydrolysis via cleavage of the ester bond to form its main degradates methomyl oxime and CO_2 (Kuhr and Dorough 1976; US EPA 1998a).

 However, environmental hydrolysis does not readily occur to a significant degree. To illustrate, Malato et al. (2002) found that methomyl solutions at either pH 2.7 or pH 5 did not significantly degrade via hydrolysis after 20 days; Tamimi et al. (2006) later verified this result at pH 6 as well. The authors of both studies concluded that hydrolysis does not occur to any significant extent in the field—at least under mild-to-strong acidic conditions.

 In the presence of Cl$^-$ (simulating the chlorination of drinking water), the rate of methomyl's hydrolytic breakdown varies; changes in pH between 7.6 and 8.9 produced half-lives differing by 30-fold (0.4–12 min; Miles and Oshiro 1990). Further investigation at pH 7.3 produced the degradation products acetic acid, methanesulfonic acid, and dichloromethylamine (all resulting from free chlorine reactions with methomyl); rates of product formation increased with increased Cl$^-$ concentrations and temperature (Miles and Oshiro 1990). Breakdown products produced at various pHs are shown in Fig. 2; at near neutral pH levels sulfoxidation occurs, whereas *N*-chlorination predominates at higher pHs (Miles and Oshiro 1990).

Photolysis
The degradation of methomyl by direct photolysis represents a minor degradation pathway; Tamimi et al. (2006) confirmed that direct photolysis occurs, but observed <4% degradation following 45 min of UV irradiation. Direct photolysis is negligible because methomyl's molar extinction coefficient is low for wavelengths higher than 290 nm (Tamimi et al. 2006); the wavelength spectrum for methomyl in aqueous

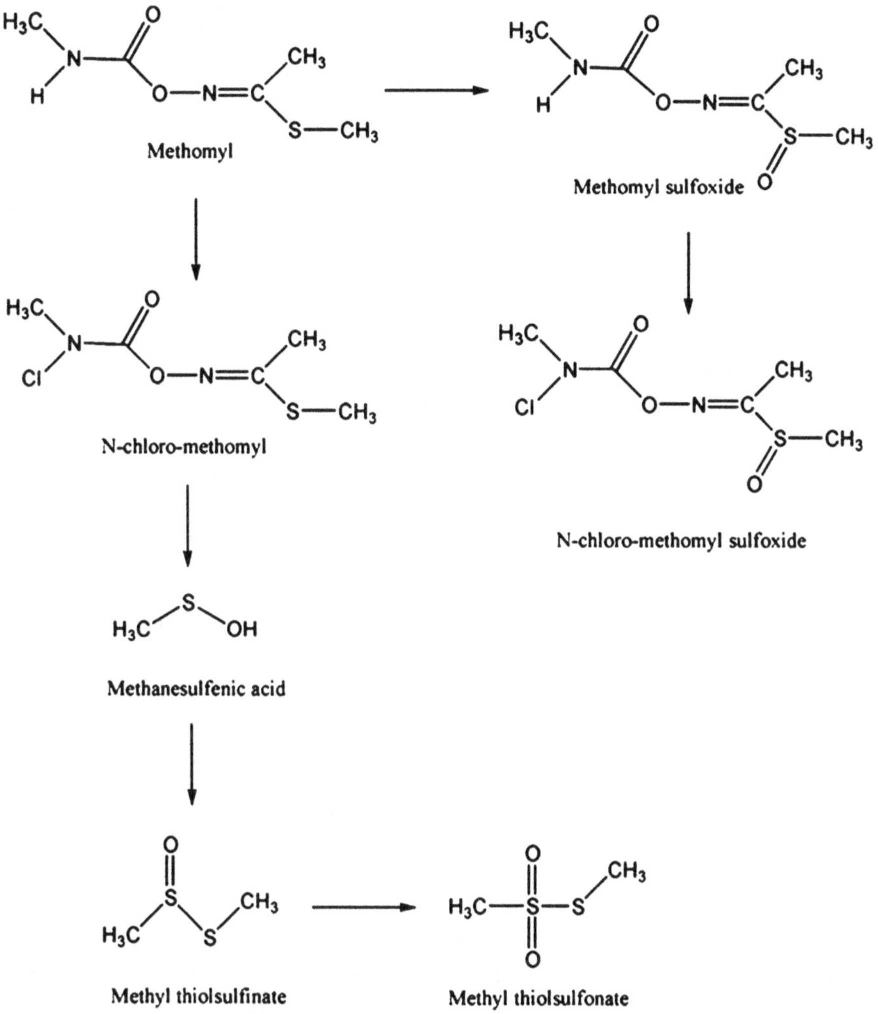

Fig. 2 Proposed reaction mechanism for methomyl under chlorinated water conditions. Additional breakdown products include acetic acid, bicarbonate, and methanesulfonic acid (adapted from Miles and Oshiro 1990)

solution is in the range of 200–300 nm, whereas the solar spectrum ranges from 300 to 600 nm (Malato et al. 2002). Tomasevic et al. (2010) investigated the influence of water quality on photolytic degradation (at 254 nm) and found that the $t_{1/2}$ of technical grade methomyl in distilled water (79.7 min; pH 5.5) was lower than that in either seawater (123 min; pH 7.9) or deionized water (97.6 min; pH 5.2). Furthermore, degradation appears to be governed by pseudo-first order kinetics, whether alone or in the presence of photosensitizers such as TiO_2 or ZnO (ZnO proved to be a better catalyst; Tomasevic et al. 2010).

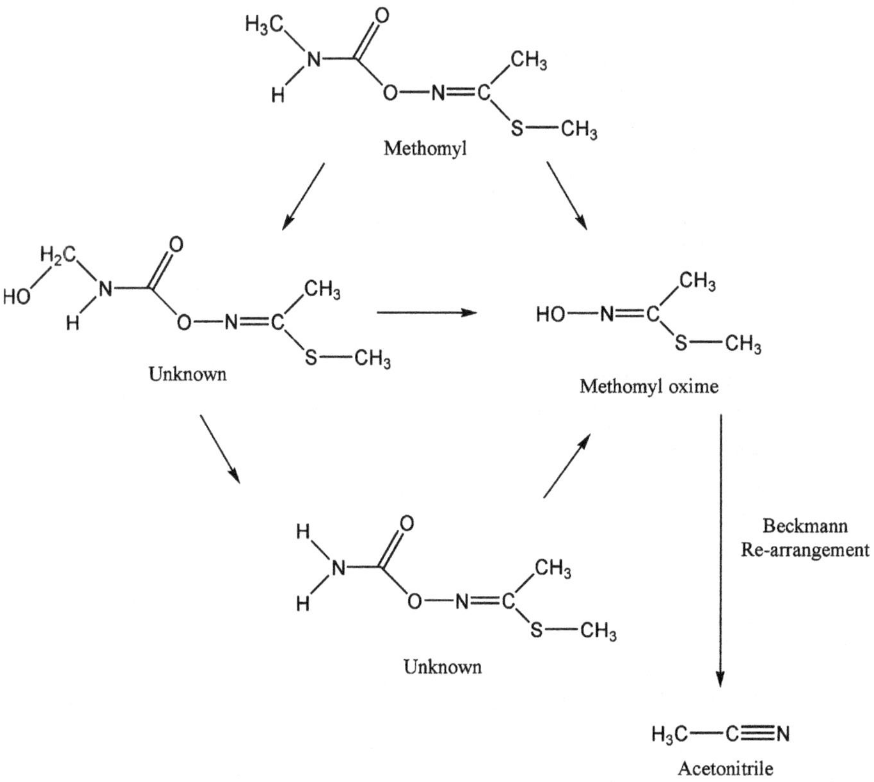

Fig. 3 Proposed photocatalytic degradation pathway for methomyl. Additional breakdown products include acetamide, acetic acid, glycolic acid, oxalic acid, formic acid, and CO_2 (adapted from Tamimi et al. 2006)

Indirect photolysis is the more efficient degradation pathway for methomyl, particularly under catalytic conditions (Fig. 3). Tamimi et al. (2008) conducted various photocatalytic experiments and found that photo-Fenton and Fenton reactions more efficiently degraded methomyl (100% and 86.1%, respectively), than did direct photolysis and $UV + H_2O_2$-catalyzed reactions (<4% and 60%, respectively). However, Mico et al. (2010) concluded that oxidation via ozonation (10.5 mg/L O_3; pH 4.5) occurs more rapidly than the photo-Fenton reaction, with complete degradation occurring within 60 min.

Chen et al. (1984) measured the photodegradative rates of various carbamate insecticides. Methomyl was degraded most rapidly when placed on a glass slide (film thickness of 0.67 μg/cm²) and irradiated in a photochemical reactor at an environmentally relevant wavelength (300 nm; 33–36°C). They also found that carbamates containing an electron-donating aliphatic group were more completely degraded than were those with an electron-withdrawing aromatic group (Chen et al. 1984).

The authors identified the photodegradative $t_{1/2}$ for methomyl to be approx. 48 h when it was applied as a thin film.

4.2 Biotic Processes

Microbial digestion of methomyl appears to be the most effective means by which it is degraded. Such degradation in two soils was investigated by Fung and Uren (1977), who employed perfusion experiments to study both soil sorption and microbial transformation; loss of the insecticide from a fine sandy loam (58%) soil was greater than that from fine sandy clay loam (38%) soil. In addition, since adsorption was minimal, the observed rapid loss of methomyl was attributed to microbial transformation (Fung and Uren 1977).

The influence of pH on the soil degradation of methomyl was studied by Harvey and Pease (1973). Under laboratory conditions (42-day exposure), methomyl applied to soil collected from the San Joaquin Valley of California (pH 7.9) degraded rapidly; upon termination of the experiment, some 45% of the radiolabeled parent compound had been converted to $^{14}CO_2$ and another 31% of the parent was retained within the soil extract (Harvey and Pease 1973).

Farre et al. (2002) evaluated the aerobic digestion of methomyl by using activated sludge populated with *Vibrio fischeri*. They found the insecticide and its major metabolite (*S*-methyl-*N*-hydroxythioacetamidate) to be completely degraded within 12 and 28 days, respectively; neither parent nor metabolite was toxic to the microbe. The gram-negative bacterium *Stenotrophomonas maltophilia* M1, isolated from contaminated irrigation sites in Egypt, was also found to significantly degrade methomyl (Mohamed 2009). Furthermore, *S. maltophilia* M1 contains a methomyl-degrading gene within plasmid PMb that is believed to be responsible for the observed digestion, and this gene is potentially transferable among other bacterial strains (Mohamed 2009).

Xu et al. (2009) recently isolated a gram-negative bacterium (*Paracoccus sp.* mdw-1) from activated sludge; methomyl appears to be used as its sole source of carbon and/or nitrogen, and warm alkaline conditions (30°C, pH 7–9) were optimal for both growth and degradation (Xu et al. 2009). Complete degradation of the insecticide (within 10-h incubation) by mdw-1 produced an unknown metabolite, which was speculated to be *S*-methyl-*N*-hydroxythioacetamidate (Xu et al. 2009). Furthermore, a microbial isolate of *Pseudomonas* spp. (EB20), cultured from a mineral slat medium (pH 7; 30°C), was observed to degrade 77% of methomyl (10 mg/L) within 2 weeks (El-Fakharany et al. 2011).

The white rot fungi isolate WR2 has also been shown to degrade methomyl and its metabolite—both in less than 42 days (Nyakundi et al. 2011). However, when combined with an additional isolate (WR9), degradation occurred more rapidly (22–25 days); accelerated degradation has generally occurred when microbe mixtures are present (Nyakundi et al. 2011).

5 Toxicology

5.1 Mode of Action

Similar to other carbamate insecticides, methomyl inhibits acetylcholinesterase (AChE), which is contained within synaptic junctions between neurons (Kuhr and Dorough 1976). When AChE is inhibited, the hydrolytic deactivation of acetylcholine (ACh) is reduced, so that it continues to stimulate the postsynaptic receptors to eventually cause nerve and/or tissue failure. In mammals, many vital functions are controlled by the peripheral nervous system, and any inhibition of these functions may lead to fatality. However, arthropods lack a peripheral nervous system (nerves outside of the central nervous system), so inhibition of AChE is not immediately fatal to them; arthropod fatality, however, may result indirectly from a secondary response caused by interrupting nerve signaling (Kuhr and Dorough 1976). Xuereb et al. (2009) exposed the amphipod *Gammarus fossarum* to various concentrations of methomyl and observed no significant mortality at 65% AChE inhibition; however, at inhibition rates of higher than 50%, significant alterations to feeding rates and locomotion were observed. Methomyl is known to cause toxicity by systemic action via direct contact or ingestion (Kuhr and Dorough 1976).

5.2 Insects

Methomyl is designed to target a broad range of insects, from immature stages to adults. Its penetration is thought to occur through the integument of the tracheal system, whereas penetration into the hemolymph is insignificant (Gerlot 1969). When radiolabeled forms of methomyl or methomyl oxime were applied to the abdomen of female face and house flies and black cutworm larvae, they both rapidly penetrated the bodies of fly species but much more slowly in cutworm larvae (Gayen and Knowles 1981). Four unknown metabolites were formed. However, the yield of $^{14}CO_2$ varied among the three species. A minimal amount of $^{14}CO_2$ was detected from insects treated with ^{14}C-methomyl oxime; this metabolite, therefore, is not considered to be a precursor for $^{14}CO_2$ formation (Gayen and Knowles 1981). Methomyl is also toxic to bulb mites (LC_{50}, 2.0 mg/L), some 15 times more so than bendiocarb. However, production of volatile degradation products such as acetonitrile and methylamine may have contributed to this toxicity (Gencsoylu et al. 1998).

Although many parasites and insects are beneficial to crops, they may also attract damaging predators that insecticidal formulations may target as well. Plapp and Bull (1978) studied the toxicity of methomyl to the tobacco budworm, its parasite *Campoletis sonorensis* and larvae of its predator *C. carnea* (a common green lacewing). The agent was highly toxic to all three species, when compared to endosulfan. In addition, toxicity data showed methomyl to be more potent towards the tobacco budworm, a pest (LC_{50}, 2.29 μg/vial), than the predator *C. carnea*

(LC$_{50}$ = 2.69 µg/vial; Plapp and Bull 1978). Hagley et al. (1981) exposed the adult parasite *Apanteles ornigis*, collected from infested apple tree leaves, to various insecticides under laboratory conditions and reported methomyl and permethrin to be equally toxic; however greater potency to adults than larvae was observed. Furthermore, methomyl was more toxic than synthetic pyrethroid insecticides to the adults of five parasitic species, when exposed to insecticide-treated filter papers for 5 days (Waddill 1978).

Davis and Kuhr (1974) investigated the toxicity of topically applied methomyl to three strains of 4th-instar cabbage looper (*Trichoplusia ni;* susceptible, DDT- and parathion-resistant). Following 48-h exposure, LD$_{50}$ values of 0.029, 0.057, and 0.34 ug/larvae, respectively, were produced. These three strains of 5th-instar cabbage loopers were injected with 2 µg of ^{14}C-methomyl and were observed to possess variable degradation rates $t_{1/2s}$ of 60, 44, and 15 min, respectively (Kuhr 1973). When compared with other tissues, metabolic activity was highest in fat body tissue homogenates. The presence of oxygen and NADPH contributed to maximum activity and results suggest that methomyl is metabolically degraded by mixed-function oxidase systems (Kuhr 1973).

Methomyl was identified as being simultaneously present with other pesticides in honey bees and brood combs in Connecticut from 1983 to 1985 (Anderson and Wojtas 1986). Pooled dead bees had measurable methomyl residues ranging from 0.04 to 3.4 mg/L; however it was less frequently detected than either methyl parathion or carbaryl, indicating that insecticide combinations may be highly detrimental to bees from additive or synergistic actions (Anderson and Wojtas 1986). However, methomyl alone is highly toxic to honey bees on contact (LD$_{50}$ <0.5 µg/bee; US EPA 1998a).

5.3 Aquatic Organisms

The bioaccumulation potential for methomyl is considered to be relatively insignificant as predicted by its log K_{ow} and water solubility values (Table 1). For example, methomyl did not significantly accumulate (<0.02 mg/L) in fish tissue following a 28-day exposure to a concentration of 0.75 mg/L (Kaplan and Sherman 1977). Although methomyl bioaccumulates only minimally, it is acutely toxic to many aquatic species. For example, it is highly toxic to *Daphnia magna* and pink shrimp (*Penaeus duorarum),* and somewhat less toxic to bluegill sunfish (*Lepomis macrochirus*) and sheepshead minnow (*Cyprinodon variegates*; Table 2). The sensitivity of *Daphnia longispina* genotypes to methomyl was compared to that of *D. magna*; the toxicity to *D. magna* (EC$_{50}$, 24.17 µg/L) was lower than the highly sensitive *D. longispina* M (EC$_{50}$, 4.71 µg/L) and *D. longispina* T (EC$_{50}$, 9.78 µg/L; Periera and Goncalves 2007).

Li et al. (2008) found that the potency of methomyl to topmouth gudgeon increased as exposure time increased; LC$_{50}$ values were 1.228 mg/L at 24 h, 0.782 mg/L at 48 h, 0.538 mg/L at 72 h, and 0.425 mg/L at 96 h, respectively.

Table 2 Toxicity of methomyl to aquatic organisms[a]

Aquatic organism	Scientific name	Test	Concentration (mg/L)
Channel Catfish	*Ictalurus punctatus*	96-h LC_{50}	0.53
Bluegill Sunfish	*Lepomis macrochirus*	96-h LC_{50}	1.05
Sheepshead Minnow	*Cyprinodon variegatus*	96-h LC_{50}	1.16
Waterflea	*Daphnia magna*	48-h EC_{50}	0.0088
Pink Shrimp	*Penaeus duorarum*	96-h LC_{50}	0.019
Mysid	*Mysidopsis bahia*	96-h LC_{50}	0.23

[a]Data from US EPA (1998a)

The insecticide significantly inhibited brain AChE activity in this species at 96 h of exposure to various concentrations. However, hepatic glutathione *S*-transferase (GST) activity showed more than a 40% decline when exposed to various concentrations for 96 h (Li et al. 2008). Yi et al. (2006) investigated the inhibition of brain AChE in both male and female carp (*Carassius auratus*). Measurement of the bimolecular carbamylation and decarbamylation rates showed that methomyl inhibited AChE in both males and females at similar rates; thus AChE sensitivity was similar between genders (Yi et al. 2006).

5.4 *Birds*

Methomyl was toxic to terrestrial game birds on an acute oral basis (ring-necked pheasant LD_{50}, 15.4 mg/kg; mallard LC_{50}, 15.9 mg/kg); however, it is slightly toxic on a subacute dietary basis (5-day LC_{50} ranges from 1,100 to 2,883 mg/L; Tomlin 2000; US EPA 1998a). Recently, contaminated corn kernels have been linked to the death of hundreds of pigeons in Medellin, Colombia; detection of methomyl in the bodies of the birds exceeded the median lethal dose for other avian species of 10–20 mg/kg (Villar et al. 2010). Pigeons exposed to methomyl suffered decreased plasma cholinesterase (ChE) levels in brain homogenates; however, ChE levels reserged, which indicates that the effects of this carbamate insecticide are reversible (Villar et al. 2010).

5.5 *Mammals*

Although methomyl targets insects, studies have shown mammals to suffer adverse effects after methomyl exposure. On an acute oral basis methomyl is highly toxic to rats, with LD_{50} values of 17–24 mg/kg and a reproductive-based No Observable Effect Concentration (NOEC) of 75 mg/L (US EPA 1998a). Erythrocytes, collected from male Wistar rats, and then exposed to methomyl underwent hemolysis, a decline in both AChE and GST activities and an increase in lipid peroxidation levels; such exposure suggests induction of oxidative damage (Mansour et al. 2009). A susceptible Chinese hamster cell line (6TG-S V79) was exposed to a log-dose range of methomyl, resulting in an LD_{50} of 959.6 µg/mL; N-nitroso methomyl was some 260 times more potent in the same test system (LD_{50} of 3.64 µg/mL; Wang et al. 1998). In another observation, significant inhibition of gap-junctional intercellular communication (GJIC) occurred at concentrations exhibiting little cytotoxicity (Wang et al. 1998).

Although methomyl poisoning in humans has not been widely studied, poisoning cases are known to have occurred. In one such case, a 60-year-old man was exposed via inhalation and transdermal absorption while spraying methomyl in his greenhouse; upon hospitalization his blood concentration was 1.6 mg/L (Tsatsakis et al. 2001). Fatalities have resulted from methomyl poisonings—both accidental and suicidal. For example, Driskell et al. (1991) reported the crash of a crop dusting plane as it sprayed methomyl onto grape seed fields. The methomyl level in the pilot's blood was 570 ± 9 ng/mL; the effects of methomyl on the pilot's nervous system were regarded to have resulted in the loss of control and crash. Miyazaki et al. (1989) reported a double suicide attempt, in which both spouses ingested methomyl powder; only one succumbed. The insecticide was measured in both the deceased spouse's serum (44 µg/g) and blood (0.2 µg/g), and an autopsy revealed multiple miliary hemorrhages in the brain—suspected to be the result of induced asphyxiation.

Human fatalities have also resulted from additive or synergistic interactions between methomyl and other chemicals. In one case, a 35-year-old male was discovered to have measurable blood concentrations of methomyl (3–8 ng/mL) and nicotine (222–733 ng/mL); both insecticides were also detected at high concentrations in the stomach. The adverse stimulatory actions of methomyl, in combination with nicotine on the nervous system resulted in death (Moriya and Hashimoto 2005).

6 Summary

The insecticide methomyl, an oxime carbamate, was first introduced in 1968 for broad spectrum control of several insect classes, including Lepidoptera, Hemiptera, Homoptera, Diptera, and Coleoptera. Like other carbamates, it inhibits AChE activity, resulting in nerve and/or tissue failure and possibly death. Considered highly toxic to insects (larval and adult stages), methomyl is thought to be metabolically degraded via mixed-function oxidase(s).

Methomyl has both a low vapor pressure and Henry's law constant; hence, volatilization is not a major dissipation route from either water or moist or dry soils. Photolysis represents a minor dissipation pathway; however, under catalytic conditions, degradation via photolysis does occur. Methomyl possesses a moderate-to-high water solubility; thus hydrolysis, under alkaline conditions, represents a major degradation pathway. Methomyl has a low-to-moderate sorption capacity to soil. Although results may vary with soil type and organic matter content, methomyl is unlikely to persist in complex soils. Methomyl is more rapidly degraded by microbes, and bacterial species have been identified that are capable of using methomyl as a carbon and/or nitrogen source. The main degradation products of methomyl from both abiotic and biotic processes are methomyl oxime, acetonitrile, and CO_2.

Methomyl is moderately to highly toxic to fishes and very highly toxic to aquatic invertebrates. Methomyl is highly toxic orally to birds and mammals. Methomyl is classed as being highly toxic to humans via oral exposures, moderately toxic via inhalation, and slightly toxic via dermal exposure. At relatively high doses, it can be fatal to humans.

Although methomyl has been widely used to treat field crops and has high water solubility, it has only infrequently been detected as a contaminant of water bodies in the USA. It is classified as a restricted-use insecticide because of its toxicity to multiple nontarget species. To prevent nontarget species toxicity or the possibility of contamination, as with all pesticides, great care should be taken when applying methomyl-containing products for agricultural, residential, or other uses.

Acknowledgments Support was provided by the Environmental Monitoring Branch of the California Department of Pesticide Regulation (CDPR), California Environmental Protection Agency, under contract No. 10-C0102. The statements and conclusions are those of the authors and not necessarily those of CDPR. The mention of commercial products, their source, or their use in connection with materials reported herein is not to be construed as actual or implied endorsement of such products. Special thanks to Kean Goh for his assistance.

References

Aktar MW, Sengupta D, Chowdhury A (2008) Degradation dynamics and persistence of quinolphos and methomyl in/on okra (Ablemoschus esculentus) fruits and cropped soil. Bull Environ Contam Toxicol 80:74–77

ACGIH, American Conference of Governmental Industrial Hygienists (1982) Threshold limit values for chemical substances in the workroom environment with intended changes for 1982. Cincinnati, OH: ACGIH

Anderson JF, Wojtas MA (1986) Honey bees (Hymenoptera: Apidae) contaminated with pesticides and polychlorinated biphenyls. J Econ Entomol 79:1200–1205

Baker LW, Fitzell DL, Seiber JN, Parker TR, Shibamoto T, Poore MW, Longley KE, Tomlin RP, Propper R, Duncan DW (1996) Ambient air concentrations of pesticides in California. Environ Sci Technol 30:1365–1368

CDPR, California Department of Pesticide Regulation's Pesticide Information Portal (2011) Pesticide Use Report (PUR) data. http://www.cdpr.ca.gov/docs/pur/purmain.htm. Accessed 7 Feb 2012

CDPR, California Department of Pesticide Regulation (2007) Summary of Pesticide Use Report Data, Indexed by Chemical. California Environmental Protection Agency, Sacramento, CA, p. 31. http://www.cdpr.ca.gov/docs/pur/pur07rep/chmrpt07.pdf. Accessed 8 Dec 2011

Carbo L, Souza V, Dores EFGC, Ribeiro ML (2008) Determination of pesticides multiresidues in shallow groundwater in a cotton-growing region of Mato Grosso, Brazil. J Braz Chem Soc 19(6):1111–1117

Chen ZM, Zablk MJ, Leavitt RA (1984) Comparative study of thin film photodegradative rates for 36 pesticides. Ind Eng Chem Prod Res Dev 23(1):5–11

Cox L, Hermosin MC, Cornejo J (1993) Adsorption of methomyl by soils of Southern Spain and soil components. Chemosphere 27(5):837–849

Davis AC, Kuhr RJ (1974) Laboratory and field evaluation of methomyl's toxicity to the cabbage looper. J Econ Entomol 67(5):681–682

Driskell WJ, Groce DF, Hill RH (1991) Methomyl in the blood of a pilot who crashed during aerial spraying. J Anal Toxicol 15:339–340

El-Fakharany II, Massoud AH, Derbalah AS, Saad Allah MS (2011) Toxicological effects of methomyl and remediation technologies of its residues in an aquatic system. J Environ Chem Ecotoxicol 3(13):332–339

Farre M, Fernandez J, Paez M, Granada L, Barba L, Gutierrez HM, Pulgarin C, Barcelo D (2002) Analysis and toxicity of methomyl and ametryn after biodegradation. Anal Bioanal Chem 373:704–709

Fung KKH, Uren NC (1977) Microbial transformation of S-methyl N-[(Methylcarbamoyl)oxy] thioacetimidate (Methomyl) in soils. J Agric Food Chem 25(4):966–969

Gayen AK, Knowles CO (1981) Penetration and fate of methomyl and its oxime metabolite in insects and twospotted spider mites. Arch Environ Contam Toxicol 10:55–67

Gencsoylu I, Liu W, Usmani A, Knowles CO (1998) Toxicological studies of the carbamates methomyl and bendiocarb in the bulb mite Rhizoglyphus echinopus (Acari: Acaridae). Exp Appl Acarol 22:157–166

Gerlot P (1969) Mode of entry of contact insecticides. J Insect Physiol 15:563–580

Hagley EAC, Pree DJ, Simpson CM (1981) Toxicity of insecticides to parasites of the spotted tentiform leafminer (Lepidoptera: Gracillariidae). Can Enr 113:899–906

Harvey J, Pease HL (1973) Decomposition of methomyl in soil. J Agr Food Chem 21(5):784–786

Hoffman RS, Capel PD, Larson SJ (2000) Comparison of pesticides in eight U.S. urban streams. Environ Toxicol Chem 19(9):2249–2258

[IPCS] International Programme on Chemical Safety (1995) Methomyl. Health and safety guide No. 97. http://www.inchem.org/documents/hsg/hsg/hsg097.htm. Accessed 8 Dec 2011

Jones RL, Hunt TW, Norris FA, Harden CF (1989) Field research studies on the movement and degradation of thiodicarb and its metabolite methomyl. J Contam Hydrol 4:359–371

Kahl G, Ingwersen J, Nutniyom P, Totrakool S, Pansombat K, Thavornyutikarn P, Steck T (2007) Micro-trench experiments on interflow and lateal pesticide transport in a sloped soil in northern Thailand. J Environ Qual 36:1205–1216

Kahl G, Ingwersen J, Nutniyom P, Totrakool S, Pansombat K, Thavornyutikarn P, Steck T (2008) Loss of pesticides from a litchi orchard to an adjacent stream in northern Thailand. Eur J Soil Sci 59:71–81

Kamrin MA, Montgomery JH (1999) Agrochemical and pesticide desk reference on CD-ROM. CRC Press, Boca Raton

Kaplan AM, Sherman H (1977) Toxicity studies with methyl N-[[(methylamino)carbonyl]oxy]-ethanimidothioate. Toxicol Appl Pharma 4:1–17

Kidd H, James DR (1991) The agrochemicals handbook, 3rd edn. Royal Society of Chemistry Information Services, Cambridge, England

Kolpin DW, Barbash JE, Gilliom RJ (2000) Pesticides in ground water of the United States, 1992–1996. Groundwater 38(6):858–863

Kuhr RJ (1973) The metabolic fate of methomyl in the cabbage looper. Pestic Biochem Phys 3:113–119

Kuhr RJ, Dorough HW (1976) Carbamate insecticides: chemistry, biochemistry and toxicology. CRC Press, Boca Raton

Leistra M, Dekker A, Van der Burg AMM (1984) Computed and measured leaching of the insecticide methomyl from greenhouse soils into water courses. Water Air Soil Poll 23:155–167

Li H, Jiang H, Gao X, Wang X, Qu W, Lin R, Chen J (2008) Acute toxicity of the pesticide methomyl on the topmouth gudgeon (Pseudorasbora parva): mortality and effects on four biomarkers. Fish Physiol Biochem 34:209–216

Mahgoub AA, El-Medany AH (2001) Evaluation of chronic exposure of the male rat reproductive system to the insecticide methomyl. Pharmacol Res 44(2):73–80

Malato S, Blanco J, Caceres J, Fernandez-Alba AR, Aguera A, Rodriguez A (2002) Photocatalytic treatment of water-soluble pesticides by photo-Fenton and TiO2 using solar energy. Catal Today 76:209–220

Mansour SA, Mossa A-TH, Heikal TM (2009) Effects of methomyl on lipid peroxidation and antioxidant enzymes in rat erythrocytes: In vitro studies. Toxicol Ind Health 25(8):557–563

Mico MM, Chourdaki S, Bacardit J, Sans C (2010) Comparison between ozonation and photo-fenton processes for pesticide methomyl removal in advanced greenhouses. Ozone- Sci Eng 32:259–264

Miles CJ, Oshiro WC (1990) Degradation of methomyl in chlorinated water. Environ Toxicol Chem 9:535–540

Miyazaki T, Yashikia M, Kojimaa T, Chikasuea F, Ochiaib A, Hidani Y (1989) Fatal and non-fatal methomyl intoxication in an attempted double suicide. Forensic Sci Int 42(3):263–270

Mohamed MS (2009) Degradation of methomyl by the novel bacterial strain *Stenotrophomonas maltophilia* M1. Electron J Biotechnol 12(4)

Moriya F, Hashimoto Y (2005) A fatal poisoning caused by methomyl and nicotine. Forensic Sci Int 49:167–170

Nyakundi WO, Magoma G, Ochora J, Nyende AB (2011) Biodegradation of diazinon and methomyl pesticides by white rot fungi from selected horticultural farms in rift valley and central Kenya. J Appl Tech Environ Sanit 1(2):107–124

Periera JL, Goncalves F (2007) Effects of food availability on the acute and chronic toxicity of the insecticide methomyl to Daphnia spp. Sci Total Environ 386:9–20

Periera JL, Antunes SC, Castro BB, Marques CR, Goncalves AMM, Goncalves F, Pereira R (2009) Toxicity evaluation of three pesticides on non-target aquatic and soil organisms: commercial formulation versus active ingredient. Ecotoxicol 18:455–463

Plapp FW, Bull DL (1978) Toxicity and selectivity of some insecticides to chrysopa carnea, a predator of the tobacco budworm. Entomol Soc America 7(3):431–434

Tamimi M, Qourzal S, Assabbane A, Chovelon J-M, Ferronatob C, Ait-Ichoua Y (2006) Photocatalytic degradation of pesticide methomyl: determination of the reaction pathway and identification of intermediate products. Photochem Photobiol Sci 5:477–482

Tamimi M, Qourzal S, Barka N, Assabbane A, Ait-Ichoua Y (2008) Methomyl degradation in aqueous solutions by Fenton's reagent and the photo-Fenton system. Sep Purif Technol 61:103–108

Tomasevic A, Mijin D, Kiss E (2010) Photochemical behavior of the insecticide methomyl under different conditions. Separ Sci Technol 45:1617–1627

Tomlin CDS (2000) The pesticide manual, 12th edn. The British Crop Protection Council, Surrey, UK, pp 620–621

Tsatsakis AM, Bertsias GK, Mammas IN, Stiakakis I, Georgopoulos DB (2001) Acute fatal poisoning by methomyl caused by inhalation and transdermal absorption. Bull Environ Contam Toxicol 66:415–420

US EPA (1989) Methomyl, PC Code 090301. http://www.epa.gov/pesticides/chem_search/cleared_reviews/csr_PC-090301_25-Oct-89_038.pdf. Accessed 8 Dec 2011

US EPA (1998a) Reregistration Eligibility Decision (RED): Methomyl. http://www.epa.gov/oppsrrd1/REDs/0028red.pdf. Accessed 8 Dec 2011

US EPA (1998b) Reregistration Eligibility Decision (RED) Facts: Methomyl. http://www.epa.gov/oppsrrd1/REDs/factsheets/0028fact.pdf. Accessed 8 Dec 2011

US EPA (2010) Problem Formulation for the Environmental Fate, Ecological Risk, Endangered Species, and Drinking Water Exposure Assessments in Support of the Registration Review of Methomyl. Docket EPA-HQ-OPP-2010-0751

Villar D, Balvin D, Giraldo C, Motas M, Olivera M (2010) Plasma and brain cholinesterase in methomyl-intoxicated free-ranging pigeons (Columba livia f. domestica). J Vet Diagn Invest 22:313–315

Waddill VH (1978) Contact toxicity of four synthetic pyrethroids and methomyl to some adult insect parasites. Florida Entomologist 61(1):27–30

Wang TC, Chiou CM, Chang YL (1998) Genetic toxicity of N-methylcarbamate insecticides and their N-nitroso derivatives. Mutagenesis 13(4):405–408

Xu JL, Wu J, Wang ZC, Wang K, Li MY, Jiang JD, He J, Li SP (2009) Isolation and characterization of a methomyl-degrading Paracoccus sp. mdw-1. Pedosphere 19(2):238–243

Xuereb B, Lefevre E, Garric J, Geffard O (2009) Acetylcholinesterase activity in *Gammarus fossarum* (Crustacea Amphipoda): Linking AChE inhibition and behavioral alteration. Aquat Toxicol 94:114–122

Yeboah PO, Kilgore WW (1984) Analysis of airborne pesticides in a commercial pesticide storage building. Bull Environ Contam Toxicol 32:629–634

Yi MQ, Liu HX, Shi XY, Liang P, Gao XW (2006) Inhibitory effects of four carbamate insecticides on acetylcholinesterase of male and female *Carassius auratus* in vitro. Comp Biochem Phys C 143:113–116

Recent Advances in Drinking Water Disinfection: Successes and Challenges

Nonhlanhla Ngwenya, Esper J. Ncube, and James Parsons

Contents

N. Ngwenya • E.J. Ncube (✉) • J. Parsons
Scientific Services Division, Rand Water, Johannesburg, 1170 2000 South Africa
e-mail: nkalebai@randwater.co.za; encube@randwater.co.za; jparsons@randwater.co.za

D.M. Whitacre (ed.), *Reviews of Environmental Contamination and Toxicology,*
Reviews of Environmental Contamination and Toxicology 222,
DOI 10.1007/978-1-4614-4717-7_4, © Springer Science+Business Media New York 2013

1   Introduction

The need for water disinfection in the developing world is undeniable. Disinfecting drinking water is critical for achieving an adequate level of removal or inactivation of pathogenic organisms that exist in raw water, for preventing recontamination of drinking water within the distribution system, and maintaining drinking water quality throughout the distribution system (USEPA 1999; AWWA 2001a; Sommer et al. 2008; WHO 2011). Waterborne diseases cause about five million human deaths per year, at least half of which are children (UNICEF 1995). Therefore, water utilities have the vital responsibility of managing water quality risks to ensure the safety and quality of water supplied to their customers. Since the beginning of the nineteenth century, inactivating and or removing pathogenic organisms (disinfection) from drinking water have been the main approaches to safeguard drinking water quality (Hrudey and Hrudey 2004). In the absence of drinking water disinfection, people are subject to falling ill from infectious diseases, caused by pathogenic bacteria, viruses, and protozoan parasites.

Chlorine-based disinfectants still play a vital role towards providing microbial-safe drinking around the world. According to a recent survey, conducted by the American Water Works Association (AWWA) (AWWA 2008), almost all surface water treatment plants in the United States of America (USA) use chlorine-based disinfectants as part of their treatment process. Moreover, it is estimated that about 98% of Western Europe's water is chlorinated (Euro Chlor 2012). Chlorine is also a widely used disinfectant in South Africa (Momba et al. 2009), Australia (CRC 2008), and China (Ye et al. 2009). Chlorination results in the effective inactivation of several pathogens and is a relatively cheap means of disinfection, and is simple to implement. However, water utilities are being driven to constantly reevaluate how they disinfect water because the environment is changing so rapidly (e.g., the increasing persistence and resistance of certain known waterborne pathogens and emergence of new waterborne pathogens, formation of disinfection by-products (DBPs) and more rapidly emerging erratic climate patterns that affect source water quality). Hence, water utilities must routinely assess the potential of alternative disinfection technologies that are sufficient to meet disinfection targets and current drinking water standards. In particular, the recent publication of the fourth edition of the WHO Guidelines for Drinking Water Quality (WHO 2011) has necessitated that water utilities review their disinfection processes.

In this review, we have aimed to provide a global overview on past and recent developments in drinking water disinfection. These developments include legislation, current and emerging technologies, and challenges and advances in drinking water risk management. In this chapter, we provide background information on the history of disinfection, the timelines over which different technologies were developed as alternatives to chlorination and also provide a synopsis of present day potable water disinfection techniques. We also present brief summaries on selected frameworks for regulating drinking water disinfection and profile the disinfection practices used around the world, and address aspects of existing and emerging potable water disinfection techniques that may help readers select a disinfectant, or improve compliance control strategies. In addition, we present an account of recent advances made in drinking water risk management that relate directly to competent strategies to provide safe drinking water to the public. We have emphasized reducing both microbial risks and risks that emanate from the formation of disinfection by-products in water distribution systems. We conclude the chapter with a summary, in which we address our view on what the future direction toward providing safe drinking water should be.

2   Background: History of Drinking Water Disinfection

The disinfection of drinking water has been hailed as one of the most important advances ever for protecting public health. Disinfection has its roots as a water treatment technique in the latter part of the nineteenth century. In the USA, chlorination was first undertaken in Louisville, Kentucky in 1896, and by the year 1908, it was fully incorporated into the drinking water treatment process (White 1999). In Europe, disinfection by water treatment first occurred in Middelkerke (Belgium) in the early 1900s and involved the use of chlorinated bleach as the disinfecting agent (White 1992). Solid calcium hypochlorite was initially used for chlorination; however, liquefied chlorine gas later became available, making large scale, continuous chlorination easier. The first permanent facilities for liquefied chlorine gas were installed at Philadelphia in 1913 and at Rye Common in England in 1917.

By World War II, chlorine disinfection had become the worldwide standard, and endured as the primary method for controlling waterborne disease (Taylor and Hong 2000). Historically, the use of chlorine for disinfection has been controversial, as many of its opponents argued that instead of disinfection, safe and protected water supplies should be used (Drown 1894). In addition, there has always been a natural aversion to using chlorine, because of the impact it has on the aesthetic qualities of drinking water. By the mid-1970s, it had been demonstrated that free chlorine reacts with natural organic constituents in water to produce chlorinated organic compounds, specifically the trihalomethanes (THMs), which pose a potential risk to human health (Rook 1974; Bellar et al. 1974). Consequently, regulators began to set limits on the amounts of THMs that could appear in finished drinking water (USEPA 1979; WHO 1996; EU 1998).

As an alternative to chlorination as a disinfection process, chloramination was discovered to be an effective means for maintaining the water quality. Chloramination produces lower amounts of THMs, while improving the taste and odor qualities of the water (AWWA 1980). The use of monochloramine as a disinfectant results in fewer chlorinated organic materials being produced. Unfortunately, little is known about the nature of these by-products except that they are more hydrophilic and larger in molecular size than are the organic halides produced from free chlorine (Health Canada 1996; USEPA 2001b). In 1916, Canada was first to use chloramines for disinfecting potable water at its principal water treatment plant in Ottawa, Ontario (Race 1918). In the USA, the first chloramination facility was established in Denver, Colorado in 1917 and the first regular application of chloramination occurred in Greenville, Tennessee, in 1926 (McAmis 1927).

By 1938, more than 400 US utilities were using monochloramine disinfection. Chloramination was employed frequently between 1929 and 1939. Soon thereafter, the use of chloramination declined as a result of the shortage of ammonia that occurred during World War II (White 1999). Chloramination was sparingly used until the 1970s, when the potentially harmful DBPs produced by chlorination were discovered. Since then, the use of chloramination as an alternative disinfection process has been steadily expanding; the promulgation of increasingly stringent regulations on DBPs have driven water utilities to switch to monochloramines (Connell et al. 2000; Mortula and Imran 2006; Cooney 2008).

Although, chlorine dioxide was produced as early as 1811, primarily for use as a bleaching agent in pulp and paper manufacture, its use in potable water disinfection was first implemented only in 1944 in Niagara Falls, New York, USA (Aieta and Berg 1986). In Europe, the large-scale use of chlorine dioxide was first introduced in 1956 in Brussels, Belgium (Block 2001). By 1977, 84 potable water treatment plants were using chlorine dioxide treatment in the USA, whereas in Europe, almost 500 plants were using chlorine dioxide (Miller et al. 1978). Chlorine dioxide produces organoleptic defects to the final water after treatment and is more biocidal than either chlorine or chloramines (USEPA 1999). Compared to chlorine disinfectants, which react with various substances via oxidation and electrophilic, substitution, chlorine dioxide reacts only by oxidation, thereby reducing THM formation in the finished water (WHO 2003a, b). Chlorine dioxide is an unstable gas that must be produced on-site by mechanical generators. Although chlorine dioxide has relatively good disinfection properties, Casson et al. (2006) suggested that its chemical instability and the complexity of on-site generation equipment represent limitations that deter use of it by public water utilities.

Apart from the introduction of stringent regulations, the early 1980s were a defining period for the potable water industry, because the protozoa, *Giardia lamblia,* and *Cryptosporidium parvum* were identified as important sources of waterborne diseases. *G. lamblia* and *C. parvum* are fairly resistant to traditional chemical disinfectants; consequently, further research and development to find alternative physical disinfection technologies were needed to remove these microorganisms (Rose et al. 1997). By the early 1990s, several effective treatment technologies for removing protozoan cysts and oocysts surfaced, including ozone, low-pressure membrane filtration, and UV light (USEPA 2001b). Ozone was discovered in 1783 by Van Marum and was named by Schonbein in 1840. In 1857, the first electric discharge ozone generator was constructed by Siemens; this device was first used commercially to disinfect potable water in 1893 at Oudshoorn, Netherlands and afterwards (1906) in France (Rakness et al. 1984).

Ozone was initially employed in the USA, in 1906, primarily to control taste and odor of water in New York City's Jerome Park Reservoir. By 1987, five US water treatment facilities were using ozone to control the taste and odor of water and to remove THM (Glaze 1987). In conventional treatment plants, potassium permanganate ($KMnO_4$) solution is added to the raw water intake, primarily to control taste and odors, to remove color, to control biological growth, and to remove iron and manganese. Potassium permanganate has a beneficial effect on disinfection, because it is a strong oxidant that reduces the needed amount of the primary disinfectant chemical (USEPA 1999). Since the 1993 outbreak of *Cryptosporidium* in Milwaukee, WI (MacKenzie et al. 1994), the use of ozone as a disinfectant has intensified. Ozone has been proven to be effective against protozoan cysts and oocysts (Kaminski 1994). Subsequently, another treatment (viz., ultraviolet radiation in a photochemical process) was demonstrated to be very effective against both *Giardia* and *Cryptosporidium* (Craik et al. 2001). Even with these new disinfection technologies, chlorine has remained the dominant method for drinking water disinfection, and similarly remains as the cornerstone of water treatment where public health is concerned.

3 Regulatory Frameworks for Potable Water Disinfection

3.1 The WHO Guidelines for Drinking Water Quality (WHO 2011)

The basic and essential requirement to ensure safe drinking water is constructing a regulatory framework comprising health-based targets, creating adequate and properly managed treatment systems, and performing independent surveillance (WHO 2011). The WHO Guidelines for Drinking Water Quality are regarded globally as the most authoritative framework on drinking water quality, and often forms the basis for other national laws and regulations (Kusumawardaningsih 2010). The WHO launched the fourth edition of its drinking water guidelines on July 4, 2011. These guidelines put greater emphasis on preventing waterborne diseases and on reducing ill-health and death from drinking contaminated water, as well as the development of water safety plans.

In the WHO guidelines, it is recommended that all surface and ground waters that are subject to fecal contamination should be adequately disinfected before being distributed for drinking purposes. To ensure the delivery of microbiologically safe

Table 1 Guideline values for chemical contamination of drinking water (WHO 2011)

Compound	Guideline value (mg/L)[a]	Remarks
Disinfectants		
Chlorine	5	Also for taste and odor control
Monochloramine	3	
Chlorine dioxide	ND[b]	
Chemical contaminants		
Nitrate (as NO_3^-)	50	The sum of the ratios of the concentrations
Nitrite (as NO_2^-)	3	as reported or detected in the sample
		of each to its guideline value should
		not exceed 1
Disinfection by-products		
Bromate	0.01	
Bromodichloromethane	0.06	
Bromoform	0.1	
Chlorate	0.7	
Chlorite	0.7	
Chloroform	0.3	
Dibromoacetonitrile	0.07	
Dibromochloromethane	0.1	
Dichloroacetate	0.05	
Dichloroacetonitrile	0.02	
Monochloroacetate	0.02	
N-Nitrosodimethylamine	0.0001	
Trichloroacetate	0.2	
2,4,6-Trichlorophenol	0.2	
Trihalomethanes		The sum of the ratio of the concentration
		of each to its respective guideline
		value should not exceed 1

[a] Guideline values for chronic health effects, unless stated otherwise
[b] Not determined

water, a residual disinfectant is recommended for use in the distribution system to provide a partial safeguard against low-level microbial contamination and growth, where necessary. For example, to be effective, WHO recommends a residual concentration of free chlorine of ≥ 0.5 mg/L be used for a contact time of at least 30 min at pH <8.0; the minimum residual concentration of free chlorine at the point of delivery should be 0.2 mg/L (Table 1). A similar guideline value for chlorine dioxide is not set, because it rapidly breaks down to chlorite; therefore, a provisional chlorite guideline value is used instead. Although maintaining a residual disinfectant in water is highly recommended, such disinfection does not offer complete protection, because the residual level may be inadequate or may be ineffective against some or all pathogen types introduced. The guidelines also include the latest guidance on what constitutes the microbial drinking water contaminants of most concern.

Table 2 Chemical and microbiological determinants for meeting the SANS 241: 2011 (SABS 2011)

	Compound	Standard
Chemical	Free chlorine	≤5 mg/L
	Monochloramine	≤3 mg/L
	Nitrate	≤11 mg/L
	Nitrite	≤0.9 mg/L
	Chloroform	≤0.3 mg/L
	Bromoform	≤0.1 mg/L
	Dibromochloromethane	≤0.1 mg/L
	Bromodichloromethane	≤0.06 mg/L
Microbiological	*E. coli* or fecal coliforms	0 per 100 mL
	Cytopathogenic viruses	0 per 10 L
	Protozoan parasites	0 per 10 L
	Total coliforms	≤10 per 100 mL
	Heterotrophic plate count	≤1,000 per mL
	Somatic coliphages	0 per 10 mL

3.2 The South African National Standard for Drinking Water Quality

In South Africa, the quality of the domestic water supply that is considered to be safe for human consumption is assured by routine monitoring for compliance against the South African National Standard for Drinking Water Quality (SANS 241). The previous rule (SANS 241:2006) featured a two-tier water quality standard. Based on the allowable time for exposure and extent of health risk, drinking water quality was categorized as Class 1 and Class 2, in which Class 1 is superior and presents minimal risk for lifetime consumption. In 2011, a revised version of the standard was published to facilitate compliance with the latest international regulatory developments for drinking water quality. Compared to the previous standard, the current version (SANS 241:2011) contains significant changes; in particular, the numerical limits for the suitability and acceptability of treated water for drinking purposes were directly derived from the WHO Guidelines for Drinking Water Quality (WHO 2011).

The current standard has two parts: SANS 241–1 and SANS 241–2; Part I is a specification of the microbiological, aesthetic, physical, and chemical numerical limits (SABS 2011). Part II prescribes how to achieve the numerical limits contained in SANS 241–1, which includes mandatory water quality risk assessment, routine monitoring, response monitoring, verification of water quality, and the requirement for a comprehensive water quality safety plan. According to this standard, water that is intended for drinking should not contain harmful concentrations of chemicals or microorganisms and should ideally have a pleasant appearance, taste, and odor (SABS 2011).

Table 2 shows the determinants of drinking water quality as specified in the SANS 241 standard. It is the ultimate responsibility of the water services institution

Table 3 Maximum residual disinfectant levels (MRDLs) set by the US EPA, based on the maximum residual disinfectant level goals (MRDLGs) (US EPA 1999)

Disinfectant	MRDLG (mg/L)	MRDL (mg/L)
Chloramine	4	4 (as annual average)
Chlorine	4	4 (as annual average)
Chlorine dioxide	0.8	0.8

Table 4 Primary drinking water regulations related to microbiological contaminants (US EPA 1999)

Compound	MCLG[a] (per 100 mL)	Remarks
Cryptosporidium	0	Water utilities to disinfect and filter their water
Giradia lambia	0	Water utilities to disinfect and filter their water
Legionella	0	Water utilities to disinfect and filter their water
Total coliform	0	No more than 5.0% positive samples per month
Viruses	0	Water utilities to disinfect and filter their water

[a]Maximum contaminant level goal

to ensure that the final water complies with this standard. Moreover, this standard is referenced in Regulation 5 (as to the quality of potable water of Section 9 vs. the standards of the Water Services Act (Act 108 of 1997).

3.3 Regulating Disinfectants in the USA

Consonant with the requirements of the Safe Drinking Water Act (SDWA) amendments of 1996, the US Environmental Protection Agency (USEPA) has developed regulations to control microbial pathogens, disinfectant residuals, and DBPs in drinking water. According to Boyd (2006), the USA imposes the most rigorous standards for protecting the public health from microbiological contaminants, because it employs outcome-based treatment standards that require a high level of effectiveness in addressing the presence in water of bacteria, viruses, and protozoa. The Safe Drinking Water Act also requires the USEPA to periodically review the national primary drinking water regulation for each contaminant and to revise the regulation, if appropriate, based on new scientific data (USEPA 1996). Tables 3, 4, and 5, respectively, show the current US primary drinking water regulations that relate to residual disinfectants, microbial contaminants, and DBPs.

The above-mentioned federal EPA regulatory values serve merely as a guide; individual states may set more stringent drinking water goals as either MRDLGs (maximum residual disinfectant level goals) or MRDLs. MRDLG are nonenforceable health goals and do not reflect the benefit of adding disinfectant to control waterborne microbial contaminants (USEPA 1999). The MRDLG is the maximum level of a disinfectant added for water treatment at which no known or anticipated health effects occur, and which allows an adequate margin of safety. In contrast, MRDLs are enforceable standards that indicate the maximum level of a disinfectant, added for water treatment that may not be exceeded, without an unacceptable possibility of

Table 5 Primary drinking water regulations related to disinfection by-products (DBPs) (US EPA 1999)

Compound	MCLG[a] (mg/L)	MCL[b] (mg/L)
Bromate	0	0.01
Chlorite	0.8	1
Haloacetic acids (HAA5)[c]	N/A	0.06
Total trihalomethanes (TTHMs)	N/A	0.08

[a]Maximum contaminant level goal
[b]Maximum contaminant level
[c]Sum of the five regulated haloacetic acids: monochloroacetic acid (MCAA), dichloroacetic acid (DCAA), trichloroacetic acid (TCAA), monobromoacetic acid (MBAA), and dibromoacetic acid (DBAA)

adverse health effects (USEPA 1999). The Surface Water Treatment Rule (SWTR) requires disinfectant residual monitoring at the entry point to the distribution system and in the distribution system. At the entry point, the disinfectant residual concentration must be not less than 0.2 mg/L for more than 4 h and must be monitored continuously (USEPA 1989). Residual disinfectant concentrations must be detected in at least 95% of the samples taken each month for 2 consecutive months.

Alternatively, herotrophic plate counts (HPCs) may be made instead of measuring disinfectant residuals. If used, and the HPC is <500 colonies/mL, the site is regarded to have the equivalent of a "detectable residual." The SWTR also seeks to improve public health protection by controlling microbial contaminants and requires removal and/or inactivation of viruses, *Legionella*, *G. lamblia* and *Cryptosporidium*. These disease-causing microbes are present at varying concentrations in most surface waters. The rule requires that treatment systems filter and disinfect surface water from supplies to reduce the occurrence of unsafe levels of these microbes (2, 3, and 4 log inactivation for *Cryptosporidium, Giardia* cysts and viruses, respectively) (USEPA 1989).

Other pertinent US M-DBP rules (Microbial and Disinfection By-products Rules), and regulations relevant to drinking water disinfection, disinfectants, and disinfection by-products include the following.

- *Stage 1 Disinfectants and Disinfection By-products Rule (Stage 1 DBP) (December 16, 1998)*: This rule is designed to reduce exposure to disinfection by-products for customers of community water systems and non-transient non-community systems, including those serving fewer than 10,000 people, in which a disinfectant is added to drinking water during any part of the treatment process (USEPA 2006a, b).
- *Stage 2 Disinfectants and Disinfection By-products Rule (Stage 2 DBP) (December 15, 2005)*: This rule builds upon earlier ones that addressed disinfection by-products to improve drinking water quality and to provide additional public health protection from disinfection by-products (USEPA 2006a, b).
- *Long Term 1 Enhanced Surface Water Treatment (LT1) Rule (January 14, 2002)*: This enhances the requirements of the 1989 Surface Water Treatment Rule (SWTR). It applies to public water systems that use surface water or ground water under the direct influence (GWUDI) of surface water and serve fewer than

10,000 people. The main purpose of the LT1 Rule is to improve public health protection by controlling microbial contaminants, particularly *Cryptosporidium* in drinking water, and to address risk trade-offs that result from the presence of disinfection by-product residues. The LT1 Rule provisions fall into four categories: (1) *Cryptosporidium* Removal—in which all systems must achieve a 2 log or 99% removal of *Cryptosporidium*; (2) Enhanced Filtration Requirements—in which all filtered systems must comply with strengthened combined filter effluent (CFE) turbidity performance requirements to assure 2 log removal of *Cryptosporidium*; (3) Microbial Inactivation Benchmarking—in which systems are required to develop a profile of microbial inactivation; and (4) Other Requirements—wherein finished water reservoirs for which construction begins 60 days after promulgation of the rule must be covered; and unfiltered systems must comply with updated watershed control requirements that add *Cryptosporidium* as a pathogen of concern. In addition, *Cryptosporidium* is included as an indicator of GWUDI (USEPA 2006a, b).

- *Long Term 2 Enhanced Surface Water Treatment (LT2) Rule (January 5, 2006)*: This rule is designed to reduce illness linked with *Cryptosporidium,* and other disease-causing microorganisms in drinking water. The rule supplements existing regulations by targeting additional *Cryptosporidium* treatment requirements in higher risk systems. This rule also contains provisions to reduce risks from uncovered finished water reservoirs and to ensure that systems maintain microbial protection when they take steps to decrease the formation of disinfection by-products that result from chemical water treatment. The rule applies to all systems that use surface or ground water that is under the direct influence of surface water (USEPA 2006a, b).

- *Groundwater Rule (GWR) (November 08, 2006)*: This rule offers improved protection against microbial pathogens in public water supply systems that use ground water sources. The rule also applies to any water supply system that mixes surface and ground water, if the ground water is added directly to the distribution system, and is provided to consumers without treatment. The rule addresses risks through a risk-targeting approach that relies on four major components: (1) Periodic sanitary surveys of ground water systems, (2) Source water monitoring to test for the presence of *E. coli, enterococci,* or coliphage in the sample, (3) Corrective actions required for any system having a significant deficiency or source water fecal contamination, and (4) Compliance monitoring to ensure that treatment technology installed to treat drinking water reliably achieves at least 99.99% (4-log) inactivation or removal of viruses (USEPA 2006a, b).

3.4   The Guidelines for Canadian Drinking Water Quality

Despite having a natural wealth of fresh water, Canada has weaker drinking water quality guidelines than those of at least one other nation or those of the World Health Organization (Boyd 2006). On behalf of the Federal-Provincial-Territorial

Table 6 Canadian guideline values for drinking water microbiological contaminants (Health Canada 2010)

Compound	Guideline value
Total coliforms	0 per 100 mL sample
Protozoa	0 per 100 mL sample
Enteric viruses	0 per 100 mL sample
Cynaobacteria toxins	0.0015 mg/L

Table 7 Canadian guideline values for chemical contaminants (Health Canada 2010)

Compound	Guideline value (mg/L)
Bromate	0.01
Chlorite	1.0
Total haloacetic acids (HAA5)[a]	0.08
Total trihalomethanes (TTHMs)	0.1

[a]Sum of the five regulated haloacetic acids: monochloroacetic acid (MCAA), dichloroacetic acid (DCAA), trichloroacetic acid (TCAA), monobromoacetic acid (MBAA), and dibromoacetic acid (DBAA)

Committee on Drinking Water (CDW) Health, Canada publishes the document entitled: The Guidelines for Canadian Drinking Water Quality (Health Canada 2010). Though these guidelines recognize that microbiological contaminants are the greatest threat to public health, and recommend filtration, there are no outcome-based standards in them for effective treatment to address the problem. Only five provinces require the filtration of surface water—Nova Scotia, Quebec, Ontario, Saskatchewan, and Alberta. Some individual communities, in their provinces and territories without mandatory filtration, do filter water on a voluntary basis, but these communities are exceptions to the rule.

Table 6 shows the guideline values for microbiological determinants as stipulated by the Guidelines for Canadian Drinking Water Quality. Although there is no stipulated numerical guideline value for HPC, protozoa or enteric viruses, it is desirable that neither human enteric viruses nor viable protozoa (e.g., *Giardia*) be detected after water disinfection. Canada's guideline for cyanobacterial toxins is 0.0015 mg/L, and this value is somewhat weaker than similar limits set by other nations; for example, the similar guideline value of 0.0013 mg/L was set by Australia. Table 7 shows a list of several current numerical guideline values that exist for chemical contaminants. These guideline values are either health-based, are listed as maximum acceptable concentrations (MAC), or were established based on operational considerations and then listed as Operational Guidance Values (OG).

3.5 The Australian Drinking Water Guidelines

The Australian Drinking Water Guidelines (ADWG) (NHMRC 2011) are designed to provide an authoritative reference for producing safe, good quality drinking water. According to the ADWG, drinking water is defined as water intended primarily for

Table 8 Australian drinking water guideline values for chemical contaminants (NHMRC 2011)

Compound	Guideline value[a] Health	Aesthetic	Remarks
Bromate	0.02		By-product of disinfection using ozone
Carbon tetrachloride	0.003		Impurity in chlorine used for disinfection
Chlorate	0.3		By-product of chlorine dioxide
Chlorine	5	0.6	Odor threshold generally 0.6 mg/L, but 0.2 mg/L for a few people
Chlorine dioxide	1	0.4	Oxidizing agent and disinfectant
Chlorite	0.8		By-product of chlorine dioxide
Chloroacetic acids			By-product of chlorination
Chloroacetic acid	0.15		
Dichloroacetic acid	0.1		
Trichloroacetic acid	0.1		
Chlorophenols			By-product of chlorination of water containing phenol or related chemicals
2-Chlorophenol	0.3	0.0001	
2,4-Dichlorophenol	0.2	0.0003	
2,4,6-Trichlorophenol	0.02	0.002	
Cyanogen chloride	0.08		By-product of chloramination
Formaldehyde	0.5		By-product of ozonation
Iodine	ND[b]	0.15	Disinfectant, taste threshold = 0.15 mg/L
Monochloramine	3	0.5	Disinfectant. Odor threshold 0.5 mg/L
Nitrate (as nitrate)	50		Guideline value will protect from methaemoglobinaemia
Nitrite (as nitrite)	3		Rapidly oxidized to nitrate (see above)
N-Nitrosodimethylamine	100 ng/L		By-product of chloramination and to a lesser extent chlorination
Silver	0.1		Concentrations generally very low. Silver and silver salts occasionally used for disinfection
Trichloroacetaldehyde (chloral hydrate)	0.02		By-product of chlorination
Total THMs	0.25		By-product of chlorination and chloramination

[a]Guideline value in mg/L, unless stated otherwise
[b]Not determined

human consumption, either directly, as supplied from the tap, or indirectly, in beverages, ice, or foods prepared with water. The ADWG are intended to provide the best evidence-based advice to professionals who manage water supplies, and in a way that the community can understand and use to participate in decision making. The guidelines are based primarily on the latest WHO recommendations and are used widely throughout Australia. The guidelines are concerned with water safety from a health viewpoint, and its aesthetic quality, i.e., its taste, color, and odor. Table 8 shows the Australian guideline values for selected chemical contaminants that are associated with disinfection.

Table 9 E Directive	Compound	Standard
health-based standards	Bromate	10 μg/L
for chemical parameters	Nitrate	50 mg/L
(EU 1998)	Nitrite	0.5 mg/L
	THMs	100 μg/L

Australia's guideline for cyanobacterial toxins is 0.0013 mg/L, which is comparable to the Canadian guideline of 0.0015 mg/L. The Australian and European guidelines also suggest that turbidity should never exceed 1 Nephelometric Turbidity Unit (NTU). The guideline values are subject to a rolling review to ensure that they are kept up to date as new knowledge develops. The ADWG stipulate that *E. coli* should not be detected in any 100 mL sample of drinking water. In the event that *E. coli* is detected, a repeat sample is required to be taken from the same site and from the immediate upstream treated sources of supply and tested for the presence of *E. coli* (or thermo-tolerant coli forms). If any of the samples are positive, then increased disinfection and a full sanitary survey should be implemented immediately.

3.6   The EU Drinking Water Directive (Council Directive 98/83/EC 1998)

The European Union (EU), currently composed of 25 member states, sets drinking water regulations for all of its member states. The European Union Council Directive 98/83/EC was adopted on November 3, 1998 to regulate the quality of water intended for human consumption (EU 1998). The Drinking Water Directive (DWD) sets standards for the most common substances (so-called parameters) that can be found in drinking water. In the DWD a total of 48 microbiological and chemical parameters must be monitored and tested regularly. In principle, WHO drinking water guidelines are used as a basis for the standards established under the Drinking Water Directive. Member States may, for a limited time, deviate from certain chemical quality standards. This process is called "derogation." Derogation can be granted, provided it does not constitute a potential danger to human health, and provided that the supply of water intended for human consumption in the area concerned cannot be maintained by any other reasonable means.

The Directive applies to all water supplies except nationally recognized mineral waters or water used as a medicinal product. The EU Directive neither specifically requires water supplies to be disinfected nor requires residuals disinfection, although the Directive suggests that disinfection be carried out when necessary. The specific parametric values for microbiological quality require that *E. coli* and enterococci be non-detected in any 100 mL water sample by using the accepted detection methods. Table 9 shows these EU Directive requirements for chemical parameters at the customer's tap. EU member states may adopt standards and monitoring requirements more stringent than those imposed by the EU Directive.

Three countries, Spain, Portugal, and the UK, require primary disinfection for all water supplies. Four countries, Austria, Denmark, France, and the Netherlands, require primary disinfection of surface water, but not for groundwater, unless necessary. No other country in the EU requires primary disinfection as a national standard. Of the 15 original EU member states, only Spain and Portugal require secondary disinfection (or residual disinfection) in distribution systems. Germany and Austria require residual disinfectants as necessary to achieve microbiological standards (no pathogens). Belgium, Finland, France, Ireland, Luxembourg, and Switzerland (not an EU member state) offer guidance on disinfectant residuals. Some European regulators monitor heterotrophic bacteria, while others do not use microorganisms as indicators of water quality (Hydes 1999).

4 Global Trends in Potable Water Disinfection

4.1 South Africa

In an effort to provide clean and safe drinking water, most drinking water treatment facilities in South Africa implement some measure of disinfection before the water is distributed. Currently, chlorine is the most widely used disinfectant in South Africa (Genthe and Kfir 1995). A recent survey by Momba et al. (2009), involving 181 small drinking water treatment plants across seven provinces of South Africa: Mpumalanga, Limpopo, North West, Free State, KwaZulu-Natal, Eastern Cape, and Western Cape also indicated that chlorination is the predominant type of disinfection used. Approximately 69% of the treatment plants use chlorine gas, about 15% use sodium hypochlorite (15%), and about 14% use calcium hypochlorite. Similarly, chlorination is one of the most commonly used disinfectants among large water treatments plants, including Rand Water. However, in some cases, chlorine-based disinfectants may simply not be sufficient for preserving the microbial quality of the water in the distribution network. Therefore, the use of chloramines as a secondary disinfectant also occurs. It is known that chloramination is practiced at the following facilities: Vaalkop water treatment plant of Magalies Water, Umzonyana water treatment plant in East London, Rand Water booster stations (Zwartkopjes, Palmiet, Eikenhof, and Mapleton), and Umgeni Water treatment plants at Hazelmere, DV Harris, and Midmar (van der Walt et al. 2002).

Despite the high operational costs, a survey conducted by Rajagopaul et al. (2008) on the use of ozone in the South African water industry indicated that the use of ozone as a pre-oxidant in the pretreatment and intermediate stages of the water treatment process train is steadily increasing. Nearly all facilities preferred chlorine as the final disinfectant over ozone due to its short half life. Examples of water utilities that have incorporated ozonation include the Wiggins water works of Umgeni Water, central water purification works at Plettenberg Bay, Midvaal Water Company, and Western Transvaal Regional Water Company (WTRWC). Ozonation at these facilities is mainly used to overcome problems associated with the oxidation of iron,

Table 10 A survey of disinfection practices in the USA (AWWA 2008)

Disinfection technology	Percentage use rate	
	1998	2007
Chlorine	98	107[a]
Chloramines	11	30
Chlorine dioxide	4	8
Ozone	2	9
UV	0	2

[a]Total percentages may be more than 100% due to the use of multiple chlorine disinfectants

manganese, and organics, as well as decolorizing the water. In addition, other added benefits of the ozonation facilities include the breakdown of taste and odor compounds, disinfection of the water, reduction in the levels of THMs after chlorination and a reduction in chlorine demand (Pieterse et al. 1993; Bauman et al. 2002; Pryor et al. 2002; MacPherson and Lombard 2006).

Momba et al. (1998) reported that ultraviolet (UV) irradiation is a disinfection method that has gained popularity in the South African potable water industry. The most important benefit of using UV disinfection is its ability to inactivate *Cryptosporidium* oocysts, which are generally resistant to the effects of most other chemical disinfectants (Bukhari et al. 1999; Shin 2000). Examples of where UV light have been employed include the current ongoing work to install a combined UV/chlorine dioxide unit for inactivating *Cryptosporidium* and *Giardia* (CSV Water 2011), and the recent installation of UV disinfection units at a Temba Water Works site of the City of Tshwane Metropolitan Municipality (PCI Africa 2005).

Recent progress in UV light disinfection of water has permitted this technology to be extended for use in treating wastewater effluent from large sewage treatment plants. The first large-scale UV plant in South Africa was installed at Pretoria's Daspoort sewage treatment plant (STP) and was successfully commissioned in July 1997. Since then, a number of UV disinfection facilities have been installed around the country, including the Potsdam wastewater works in the City of Cape Town (Cannon et al. 2008). Although UV irradiation is a good disinfection process for killing pathogens and other organisms that are resistant to chlorination, it does not offer residual protection. Consequently, UV disinfection, as for ozone, is not the preferred method for primary disinfection of potable water. This technology is, however, well suited for point source use (USEPA 2001b).

4.2 United States of America

Most US water treatment plants disinfect water prior to distribution. In 2007, the AWWA Disinfection Systems Committee conducted its fourth survey of drinking water utility disinfection practices (AWWA 2008). Table 10 displays a breakdown of the chemical usage that is based on the AWWA Disinfection Systems Committee survey's data. The most commonly used disinfectants/oxidants are chlorine, chlorine dioxide, chloramines, ozone, and UV light (AWWA 2008). The table shows that chlorine is still the predominant disinfectant/oxidant, with a 107% rate of use by water treatment plant systems in the USA, indicating a 9% increase in the number of facilities using chlorine for disinfection from 1998 to 2007. Although chlorine is desirable and effective in treating water to meet certain regulatory standards, its use has been associated with the presence of undesirable DBPs in the distribution system (USEPA 1979). With the promulgation of the Long-Term 2 Enhanced Surface Treatment Rule (LT2ESWTR) and the Stage 2 Disinfectant/Disinfection By-product Rule, the USEPA has adopted more stringent Maximum Contamination Levels (MCLs) of 80 µg/L for total trihalomethanes (TTHMs) and 60 µg/L for the five haloacetic acids (HAAs) (USEPA 2006a, b). Mukiibi and Sherwin (2011) indicated that the use of chlorination without a pre-oxidation step is therefore expected to decline. The 2007 survey also indicated a 19% increase in the use of chloramines. Since the late 1990s, many water utilities in the USA have switched to chloramination for secondary disinfection because it produces fewer disinfection by-products and does not form the specific by-products of concern that are associated with chlorine (USEPA 1979). Mukiibi and Sherwin (2011) stated that they expected the use of chloramines as a secondary disinfectant in the USA to increase to about 50% by 2013 from the promulgation of the Stage 2 DBPR.

The use of chlorine dioxide increased by 4%, whereas the use of ozone and (UV light as disinfection agents increased by 7% and 2%, respectively (AWWA 2008). Increases in ozone and UV are a result of sequential disinfection schemes, such as UV/combined chlorine and ozone/combined chlorine, which are also being considered by many drinking water utilities as the inactivation component of their multiple-barrier water treatment approach, because, using both UV and ozone are more effective for controlling *C. parvum* oocysts than using chlorine alone (Shannon et al. 2008).

4.3 Canada

Similarly, most water treatment plants in Canada use chlorine for both primary and secondary disinfection purposes (Health Canada 2006). For example, more than 90% of treatment plants in Newfoundland use chlorine as the primary disinfectant (DOE 2006). In another evaluation, the 2001–2004 Drinking Water Surveillance Program (DWSP) results showed that 165 of 179 (92%) treatment plants in Ontario use chlorination as their primary disinfectant (MOE 2006). Data obtained in a 2005 survey from 3,590 drinking water facilities located in nine provinces and territories indicated that sodium hypochlorite is the most common disinfectant used in 78% of plants, whereas 19% used chlorine gas, 1.4% used calcium hypochlorite, and less than 0.5% applied alternative disinfectants (Health Canada 2009). Some water utilities use monochloramine, for example, as a secondary disinfectant (Health Canada 1996).

4.4 Australia

Compared to other disinfectants, chlorine is most widely used for drinking water in Australia (The Cooperative Research Centre (CRC) for Water Quality and Treatment 2008). It is used in most Australian capital cities and by many smaller water supplies. The majority of users employ chlorine because it is relatively inexpensive, easy to use, and effective at low dosages against a wide range of infectious microorganisms and can protect water within the pipe system (Hunterwater 2009). Other widely used chemical disinfection systems in Australia are chloramines, ozone, and UV radiation. The disinfectant type/s chosen and their effectiveness depends mainly on the following variables: the nature and concentration of the disinfecting agent, type of microorganisms present, disinfectant contact time (including size of distribution network), mixing between the disinfectant and water, degree to which the microorganisms are protected by adsorption to, or inclusion in, solid particles and attachment to surfaces of pipes or fittings, the level of competing inorganic and organic reactants and water turbidity, temperature and pH (NHMRC and NRMMC 2004).

4.5 Europe

In Europe, groundwater plays a crucial role in providing water for domestic and recreational purposes, and about 75% of EU member states depend on groundwater for their water supply. In Europe, groundwater is seldom disinfected, because it is protected against microbiological contamination by legal regulations and policies, such as the new Groundwater Directive (EU 2006). This Directive provides a groundwater protection policy, in which anthropogenic activities along groundwater collection zones is prohibited. In addition, the water is abstracted by hygienic means and the treatment and storage facilities are covered and protected (European Commission 2008). Currently, the UK is one of few countries where drinking water regulations require that water for public consumption should not be supplied from any source unless it has been disinfected. Chlorine is the most widely applied primary disinfectant in Europe, followed by ozone and chlorine dioxide (Smeets et al. 2006). However, since 2006, chlorine disinfection is not used in the Netherlands. For primary disinfection in direct treatment systems (without infiltration or river bank infiltration), UV light is used, either by itself or in combination with peroxide, and in some instances ozone is used (Smeets et al. 2009).

The UK is one of few European countries that use chloramines for residual disinfection in the distribution network and for minimizing the formation of disinfection by-products. Finland, Spain, and Sweden also use chloramines for disinfection but not on a regular basis. France mainly uses ozone, while Italy and Germany use either ozone or chlorine dioxide as a primary oxidant and disinfectant (Lenntech 2011). The current approaches to secondary disinfection in Europe are influenced by the wide diversity of water resources and supply infrastructures, as well as disinfection philosophy. For example, in Western Europe, when post-disinfection occurs, chlorine dioxide is usually the agent that is applied (MWH 2005).

5 Current Issues in Potable Water Disinfection

5.1 New Perspectives on Microbiological Drinking Water Quality Indicators

Monitoring drinking water for indicator and emerging pathogens is an important aspect toward protecting public health. Monitoring programs aim to protect consumers from illness due to pathogenic organisms such as bacteria, viruses, and protozoa and thus to prevent drinking water-related illness outbreaks. For the past century, the potential presence of pathogenic microorganisms in drinking water has been evaluated by analyzing finished drinking water for fecal pollution indicators (WHO 2003a, b). However, it has recently been shown that the absence or presence of coliform bacteria in the finished water does not adequately reflect the presence or absence of other pathogenic microorganisms, e.g., viruses and protozoa (Jacangelo et al. 2003). Similarly, the HPC test does not reflect the pathogenicity of the distribution system microbial populations.

The main purpose of the HPC test is to merely provide information on treatment efficiency, extent of after-growth in distribution networks and adequacy of disinfectant residuals (SABS 2011). According to the WHO (WHO 2003a, b), there is no clear-cut evidence, either from epidemiological studies or from correlation with occurrence of waterborne pathogens, that heterotrophic bacteria pose a public health risk, particularly when they are ingested by healthy people via drinking water. Therefore, the current WHO Guidelines for Drinking Water Quality (WHO 2011) and the majority of international drinking water regulations do not provide a numerical standard or guideline value for the HPC; rather, they state that the final water quality should meet the set microbiological guideline values.

Despite the general perception that HPC does not reflect a meaningful health risk, there has been a dramatic increase in infections caused by microorganisms, including certain heterotrophic microorganisms that are found in drinking water (Huang et al. 2002). In a joint study by the University of Pretoria and Rand Water, preliminary evidence that drinking water may contain potentially pathogenic heterotrophic microorganisms, including bacteria belonging to the following genera: *Acinetobacter, Aeromonas, Aureobacterium, Bacillus, Chryseobacterium, Corynebacterium, Klebsiella, Moraxella, Pseudomonas, Staphylococcus, Tsukamurella* and *Vibrio*, was presented (Pavlov et al. 2004). However, the actual organisms detected by HPC tests vary widely among locations and among consecutive samples (Bartram et al. 2003). The proposed SANS241:2011 standard specifies an HPC limit of 1,000 counts/mL for drinking water (SABS 2011). Although this value is in line with the approach that HPCs are an operational tool for treatment efficiency, it is also an indication of the microbial quality of water in the distribution systems. Generally, the numbers of HPC organisms can be reduced significantly by disinfection practices, such as chlorination, ozonation, and UV light irradiation (WHO 2011).

Apart from HPC, there is a need for alternative indicators, which should ideally cover both the occurrence of surrogate microorganisms (index function) and their behavior (indicator function) (Ashbolt et al. 2001). Based on the purpose they serve, indicator organisms can be categorized into three major groups: process microbial indicators, fecal indicators, and index and model organisms. Process indicators are a group of microorganisms that demonstrate the efficacy of a process (e.g., the treatment process). Fecal indicators reflect the presence of fecal contamination, and thus only infer that pathogens may be present. Index and model organisms include a group or species indicative of pathogenic presence and behavior, respectively. Current drinking water regulations, such as the revised WHO Guidelines for Drinking Water Quality (WHO 2011) and SANS 241:2011 standard (SABS 2011), have both included several bacteriophage groups as alternative indicators for the presence of pathogenic viruses.

Other microbial contaminants such as *Pseudomonas, Proteus mirabilis*, and *Aeromonas* spp. are capable of growth under low nutrient conditions, similar to those found in water distribution systems, and therefore should be proposed as additional indicators of distribution system integrity. Their occurrence in water suggests inadequate chlorination and potential biofilm formation (WHO 2011). In addition, opportunistic pathogens, such as *Acinetobacter, Enterobacter, Klebsiella, Pseudomonas*, which are potentially pathogenic in persons with weakened immune systems or in other susceptible subpopulations, such as burn patients, should be monitored (Post et al. 2011). Essentially, the continued outbreaks of waterborne diseases demonstrate that pathogenic organisms in drinking water still pose a risk to public health. Although microbial risks can be reduced by using chemical disinfectants, they pose their own potential health risks by forming DBPs. Notwithstanding, priority must be given to protect public health against microbial risks, because the risk they pose is acute, as opposed to the chronic risks of cancer or reproductive effects posed by residual chemicals such as DBPs (WHO 2011).

5.2 Current and Emerging Drinking Water Pathogens of Concern

Currently, infectious agents that cause waterborne diseases include a variety of helminths, protozoa, fungi, bacteria, rickettsiae, viruses, and prions (WHO 2003a, b). Although some infectious agents have been eradicated or diminished, new ones continue to emerge that present their own challenges. According to the Centre of Disease Control and Prevention (Ewald 1996), emerging pathogens are those that have increased the incidence of human disease over the past two decades or threaten to increase the disease incidence in the near future. Emerging pathogens may include (1) a new pathogen created from the evolution of an existing organism; (2) a known pathogen spreading to a new geographic area or human population; (3) a previously unrecognized pathogen that appears in areas undergoing ecologic transformation;

and (4) previously controlled infections that reemerge as a result of increased antimicrobial resistance or breakdowns in public health measures.

Recently, scientists have identified several emerging waterborne pathogens. Among these are the following: new enteric viruses (e.g., noroviruses), *Legionella*, *Mycobacterium avium* complex, *Aeromonas hydrophila*, *Helicobacter pylori*, *Yersinia enterocolitica*, *Pseudomonas aeruginosa*, *Proteus mirabilis*, and microsporidia (Szewzyk et al. 2000). Among these emerging pathogens, viruses, and prions are of particular concern, and account for nearly half of all emerging pathogens in the last two to three decades (USEPA 2002).

5.2.1 Bacterial Waterborne Pathogens

Bacterial waterborne infections remain one major cause of human morbidity and mortality worldwide (WHO 2003a). The most important bacterial agents causing infections or epidemics through drinking water contamination include *Campylobacter* spp., pathogenic *Escherichia coli* (e.g., *E. coli* O157: H7), *Salmonella* spp., *Shigella* spp., *Vibrio cholerae*, and *Yersinia enterocolitica* (Jesperson 2004). Between 1993 and 2006, species of enteric bacteria in the genera *Escherichia*, *Shigella*, *Salmonella*, *Plesiomonas*, and *Yersinia* were identified as the main causative agents in 26 waterborne disease outbreaks in the USA (Yoon and Hovde 2008). Although all of these pathogens can cause gastroenteritis, infection with enterohemorrhagic *E. coli* O157:H7 can have severe outcomes. In vulnerable persons, particularly children and the elderly, this infectious agent can progress to hemolytic-uremic syndrome, whereby kidney failure results in serious illness or death, especially in developing countries where the health care system is poor (WHO 2003a; Ashbolt 2004). For example, mortality of up to 22% was reported for waterborne diseases outbreaks caused by the pathogens *V. cholerae* and *E. coli* O157:H7 (Hunter 1997). The most recent outbreak of *E. coli* O157:H7 occurred in Walkerton, Ontario, Canada, in 2000 and resulted in six deaths and more than 2,300 cases (Bruce-Grey-Owen Sound Health Unit 2000).

Campylobacter enteritis in man is caused mainly by *Campylobacter jejuni* or *C. coli*, which are zoonotic and are carried by wild and domestic animals, especially by birds and poultry. Moreover, the infective dose of *Campylobacter* is low, below 1,000 organisms (Blaser 1997). Unfortunately, disease caused by this organism is relatively common, with about 19 outbreak cases having occurred between 1993 and 2006. Both enteric bacteria and *Campylobacter* spp. are amenable to conventional water treatment and disinfection. Generally, most known drinking water bacterial pathogens are sensitive to inactivation by conventional chemical disinfectants, such chlorine and chlorine dioxide (Junli et al. 1997). Haas (1999) argued that had chlorination been implemented in distribution systems, both a 1993 *Salmonella* outbreak caused by animal waste introduced to a distribution system reservoir and a 1989 *E. coli* O157:H7 outbreak could have been prevented. However, certain bacteria have been reported to be highly resistant to free chlorination; an example is spore-forming bacteria such as *Bacillus* or *Clostridium* (WHO 2011).

The *Legionella* bacterium is another example of an emerging pathogen. *Legionella* are parasites of protozoa and are naturally occurring bacteria; they are widely distributed in fresh water, including groundwater and wet soil (Riffard et al. 2004). Altogether, 26 species of *Legionella* have been identified, and seven of these are etiologic agents for Legionnaires' disease (AWWA 1990). In the environment, *Legionella* may grow in the presence of algae and cyanobacteria (Fliermans 1996; WHO 2002a, b). Small numbers of these organisms have been found in distribution and plumbing system biofilms (Rogers and Keevil 1992). The primary human exposure route is thought to be via inhalation of water aerosols containing high concentrations of *Legionella*, such as during showering. *Legionella* bacteria cause legionellosis that is characterized as either a self-limiting febrile illness called Pontiac fever or a serious type of pneumonia called Legionnaires' disease; the symptoms are indistinguishable from pneumococcal pneumonia (Fields et al. 2002).

Legionella has produced 26 disease outbreaks (156 cases and 12 deaths) following exposure to water intended for drinking, during the period 2001 through 2006 (Blackburn et al. 2004; Liang et al. 2006; Yoder et al. 2008). WHO (2007) recommends taking immediate action when samples show concentrations of *Legionella* that exceed 0.1 cfu/mL or when heterotrophic bacteria counts exceed 10,000 cfu/mL. Small numbers of these organisms can survive chlorination, particularly those embedded within biofilms. Hence, alternative disinfection procedures, such as ozonation, may be necessary to inactivate them (Thomas et al. 2004; Loret et al. 2005).

Another emerging bacterial disease-causing agent of concern belongs to the Mycobacteria group. *Mycobacteria* are free-living bacteria, occur naturally in water and soil, and have been isolated from water distribution systems. As for *Legionella*, *Mycobacteria* have been shown to survive within protozoan species (Vaerewijck et al. 2005; Mura et al. 2006; Pagnier et al. 2009). Environmental *Mycobacteria*, including the *Mycobacterium avium* complex (*M. avium* and *M. intracellulare*; MAC) that were found in biofilms of water distribution systems, have been reported to cause disease invulnerable subpopulations, and thus, may be regarded as opportunistic pathogens (Pedley et al. 2004; Vaerewijck et al. 2005).

Although chlorine has excellent bactericidal properties, a study by Norton and LeChevallier (2000) showed that nearly all bacteria surviving chlorine disinfection were Gram positive, acid-fast and partially acid-fast bacteria, including species such as *Mycobacterium* and *Nocardia*. *M. avium*, in particular, is ubiquitous in biofilms within water distribution systems around the world, and has a remarkable resistance to chlorine at high pH (pH > 8) and low water temperatures (LeChevallier 2006; Shin et al. 2008). Many atypical *Mycobacterium* spp. have been detected in well-operated and maintained drinking water supplies that have levels of HPC <500 cfu/mL and total chlorine residuals up to 2.8 mg/L (WHO 2011). Other examples of opportunistic emerging pathogens include the following: *Acinetobacter*, *Enterobacter*, *Klebsiella*, *Serratia*, *Aeromonas*, and *Pseudomonas* (AWWA 2006). Most of these bacteria are naturally present in the environment and occur in the biofilms of various water distribution systems around the world (Regan et al. 2003; Beech and Sunner 2004; Camper 2004; Emtiazi et al. 2004). They have been reported to cause disease in vulnerable subpopulations, such as the elderly or the very young, patients with burns or extensive wounds, those undergoing immunosuppressive therapy or those with acquired immune deficiency syndrome (AIDS) (Szewzyk et al. 2000). If the people belonging to these groups use water that contains sufficient numbers of these organisms, various infections of the skin and the mucous membranes of the eye, ear, nose, and throat will result (WHO 2011).

The presence of cyanobacteria (blue-green algae) in drinking water supply systems is another growing health concern. Although cyanobacteria are not pathogens, their excessive growth (blooms) in source waters may release undesirable metabolites that are difficult and expensive to treat (Chorus and Bartram 1999). Cyanobacterial blooms result from several factors that include high temperatures, direct sunshine, high levels of nutrients in water, and low flows. Metabolites of major concern that are produced affect the taste and odor compounds, particularly 2-methyl isoborneol (MIB) and geosmin, and a range of toxic compounds collectively known as algal toxins or cyanotoxins (Global Research Coalition 2009). Cyanotoxins are produced by several free-floating, toxin-producing strains, and when ingested through drinking water, can damage the liver, kidneys, and the nervous and gastrointestinal systems. Some of these toxins also have cancer-promoting effects, and exposure to water contaminated by cyanobacteria may cause eye irritation and a skin rash when showering or bathing (Haider et al. 2003).

The most common cyanobacterial hepatotoxins are Microcystins, and they are produced mainly by cyanobacteria belonging to the genera *Microcystis*, *Anabaena*, *Planktothrix*, and *Nostoc* (Spoof 2004). About 76 different microcystin analogues have been identified in natural blooms and laboratory cultures of cyanobacteria, and the most common variant is microcystin LR (MC-LR). Other microtoxin variants that have been identified in natural water samples include MC-LA, MC-RR, and MC-YR (Yoo et al. 1995; Falconer et al. 1999; Spoof et al. 2003). MC-LR is the most toxic microcystin and has an LD_{50} value of 50 µg/kg (Dawson 1998). Consequently, the WHO has set a guideline limit value of 1 mg/L for MC-LR and is proposing the same concentration for the liver toxin, cylindrospermopsin (CYL) in drinking water (WHO 2011). Currently, there is insufficient information to set acceptable levels for any of the other microcystin toxins (e.g., microcystin-LA, -YR, and -YM) or for any of the other hepatotoxins or neurotoxins that could be present (Rodriguez et al. 2007).

MC-LR can be eliminated from natural waters by applying the oxidants and disinfectants typically used in water treatment plants. Rositano et al. (1998, 2001) have shown that MC-LR is readily oxidized to nontoxic degradation products under appropriate conditions. Nicholson et al. (1994) reported that a chlorine residual of 0.5 mg/L after 30 min contact time at pH < 8 was effective in destroying microcystin-LR, whereas inadequate chlorine doses and higher pH caused negative results. Chlorination can be used to degrade microcystins; however, it should not be considered as the sole remedial measure. Rather, it should be an option to reduce the concentration of some cyanotoxins. To better manage cyanotoxins during drinking water treatment, chlorination must be integrated in a multi-barrier approach that includes adsorption on activated carbon (Merel et al. 2010). Monochloramine, which can be formed during chlorination of ammonia-containing waters, is not capable of oxidizing microcystins (Acero et al. 2005). Westrick and coworkers (2010) suggest that the implementation of innovative technologies, such as UV disinfection and membrane filtration, may greatly improve cyanotoxin removal and inactivation efficiencies.

5.2.2   Viral Waterborne Pathogens

Viruses are microorganisms that are composed of the genetic material deoxyribonucleic acid (DNA) or ribonucleic acid (RNA), along with a protective protein coat (i.e., single, double, or partially double stranded). All viruses are obligate parasites and are unable to carry out any form of metabolism; their replication is therefore completely dependent upon the availability of host cells. Viruses are typically 0.01–0.1 μm in size and are very species-specific with respect to infection. Typically, viruses attack only one type of host. Viruses (e.g., adenoviruses, enteroviruses, hepatitis A, B, and E viruses, noroviruses, sapoviruses, and rotaviruses) have been identified as pathogenic humans and can cause acute GI disease, although some cause more severe illnesses. Most viral infections occur through fecal-oral route, are associated with a wide range of both serious less serious illnesses that include conjunctivitis, mouth and throat sores, sharp abdominal pain, rashes, fever, and respiratory and GI illnesses. No disease outbreaks from consuming drinking water have been recorded in the USA since 1993, although recreational waterborne outbreaks have occurred (Post et al. 2011).

Enteric viruses are generally more resistant to free chlorine than are enteric bacteria, with CT values for 99% inactivation ranging from ≈2 to ≥30 mg min/L (WHO 2011). Chlorination effectively inactivates viruses if the turbidity of the water is less than or equal to 1 NTU. Viruses associated with cellular debris or organic particles may require high levels of disinfection due to the protective nature of the particle surface (Hoff and Akin 1986; Hoff 1992).

5.2.3   Protozoan Waterborne Pathogens

Protozoa are single-cell eucaryotic microorganisms that utilize bacteria and other organisms for food and do not possess cell walls. Most protozoa are free-living in nature and are encountered in water; however, several species are parasitic and live on or in host organisms. Among parasitic protozoa, several species may be transmitted to humans through the drinking water route. These are *Entamoeba histolytica*, *Cryptosporidium* (primarily *C. hominis* and *C. parvum* cattle genotype), *Giardia intestinalis*, *Toxoplasma gondii*, *Balantidium coli*, *Cyclospora cayetanensis*, Microsporidia, *Isospora belli*, *Naegleria fowleri*, and *Acanthamoeba* sp. Adam (2001) and Hunter and Syed (2001) reported that *G. lamblia* and *Cryptosporidium* species are the most common waterborne pathogens that induce human gastroenteritis. *G. lamblia* is the second leading cause of drinking water disease outbreaks in the USA after *Legionella*. It is estimated that about 25 outbreaks occurred between 1993 and 2006.

Giardia cysts are relatively large (8–14 μm) and can be removed via filtration by using diatomaceous earth, granular media, or membranes (Corona-Vasquez et al. 2002). *Giardia* cysts are relatively resistant to chlorine treatment, especially at higher pH and low temperatures. Similarly, the occurrence of the environmentally resistant thick-walled oocyst stage of *Cryptosporidium* has created worldwide concern because of its resistance to disinfection with chlorine concentrations that are typically applied in drinking water treatment plants (2–6 mg/L) (Rochelle et al. 2002). Controlling protozoan oocysts remains a major challenge for drinking water utilities because they are not inactivated by chlorination, the most widely used disinfection method in the world (Corona-Vasquez et al. 2002).

The infectious dose of *Cryptosporidium* oocysts in humans was estimated to be as low as 30 oocysts (Du Pont et al. 1995). To date, more than 160 waterborne outbreaks of cryptosporidiosis have been reported globally, with the greatest incidence being recorded in the USA and the UK (Craun et al. 2002). Exposure to this organism normally occurs via ingestion of fecal-contaminated water, although direct fecal–oral contact (including touching contaminated objects) or intake of contaminated food may also occur.

5.2.4 Fungal Waterborne Pathogens

According to a report by the Department for Environment Food and Rural Affairs, UK (2011), a variety of different fungi have been identified to exist in drinking water distribution systems of many countries around the world. Fungi enter drinking water distribution systems by treatment breakthrough, deficiencies in stored water facilities, cross pipe connections, mains breaks and intrusions, and during mains installation and maintenance. Once introduced, fungal species may become established on inner pipe surfaces, including interaction and reaction with sealings and coatings, and biofilms within distribution systems, or can be suspended in the water. Currently, only 500 fungi species have been linked to human disease, of which 100 may cause disease in otherwise healthy individuals (Richardson and Warnock 2003).

The most problematic fungal species are *Candida* spp. (especially *C. albicans*), *Aspergillus* spp. (particularly *A. fumigatus*), and *Cryptococcus neoformans* (Paterson et al. 2009; Pfaller et al. 2006). Hageskal et al. (2006) stated that, although healthy individuals may suffer from superficial or localized fungal infections caused by these taxa, there is little evidence that their pathogenicity arises from the consumption of contaminated drinking water. However, more severe invasive fungal infections have been reported in individuals with immune deficiency (e.g., HIV/AIDS, chemotherapy, immunosuppressive therapy following transplants, or other underlying health conditions, such as cystic fibrosis or diabetes mellitus). Such invasive infections carry a high mortality rate that is estimated to be between 50 and 100%, depending on the pathogenic species involved.

Pathogenic fungal species have also been linked to allergic disease, by worsening of asthma symptoms, or by enhancing hypersensitivity pneumonitis and skin irritation.

Certain fungi, including *Penicillium* spp., *Aspergillus* spp., *Fusariam* spp., and *Claviceps* spp., are known to produce mycotoxins such as patulin, aflatoxins, and zearalenone. It is thought that concentrations of mycotoxins in drinking water remain low from the effects of dilution; fortunately, to date, there have been no reports of disease caused by mycotoxins in drinking water. However, fungi produce secondary metabolites that have been accused of altering the taste and odor of drinking water. It is thought that the threshold level for numbers of fungi that can cause organoleptic problems in water may be ~102–103 cfu/L. Although taste and odor do not necessarily imply a human health risk, they are often perceived to do so by the consumer (DEFRA 2011). Currently, there are only a few regulations designed to control levels of fungi in drinking water. Among the countries known to limit fungal numbers is Sweden. Sweden limits the concentration of microfungi in drinking water to 100 cfu/100 mL. This limitation applies at the point of water use, and therefore takes into account fungi that enter the system through secondary contamination pathways (National Food Administration 2001). Fungi can be easily removed by using the conventional drinking water treatment processes, such as coagulation–sedimentation–filtration (sand or granular activated carbon), in which up to 90% fungi are removed (Niemi et al. 1982; Kelley et al. 2001). Kelley et al. (1997) reported that an initial free chlorine concentration of 1.3 mg/L inactivated the following fungi: 99.36% (*Trichoderma harzianum*) after 60 min, 98.11% (*Epicoccum nigrum*) after 40 min, and 97.65% (*Aspergillus niger*) after 10 min. Ozone, by contrast, achieved 99% inactivation after 18 and 5 s at 0.02 and 1 mg/L ozone, respectively.

5.3 Recent Advances in Drinking Water Disinfection

Disinfection technologies such as chlorination, chloramination, ozonation, UV light, and chlorine dioxide have made drastic improvements in drinking water safety (USEPA 1999). In recent decades, the safety and applicability of such disinfectant technologies have been called into question. Although these technologies are generally effective against microbes, their use may produce harmful DBPs, when certain organic compounds are present in the water. Furthermore, the increased resistance of some pathogens, such as Cryptosporidium, *Giardia*, and viruses, to conventional chemical disinfection requires the application of extremely high disinfectant dosages that aggravate DBP formation and elevate operational costs. As a result, the effort to provide safe, appropriate, effective, and affordable drinking water disinfection alternatives continues, as new and old technologies are evaluated.

There are recent examples of water disinfection being achieved by using mixed disinfectants. In particular, electrochemically produced mixed oxidants that are generated in situ by electrolysis of brine have been shown to achieve considerable disinfection efficiency against a number of microorganisms (Casteel et al. 2000). During electrolysis, chlorine is produced as the primary reaction product, and ozone and chlorine dioxide are produced as secondary reaction products (Patermarakis and Fountoukidis 1990; MIOX 1995). Another approach that has gained popularity involves using ozone or chlorine dioxide as the primary disinfectants, followed by using chlorine as the secondary disinfectant; this approach is known as sequential disinfection. This is an effective method for treating various pathogens (Driedger et al. 2000; Corona-Vasquez et al. 2002; Cho et al. 2003). Other emerging drinking water disinfection methods include using solar (UV) radiation, ultrasonic treatment, membrane technology, silver electrochemistry, bromine, titanium dioxide, and potassium permanganate. Unfortunately, some of these methods (e.g., treatment with bromine, titanium dioxide, and potassium permanganate) have limited utility because of low efficacy, high costs and/or the fact that they often produce toxic DBPs (Batterman et al. 2000). Some of the more utilitarian of newer disinfection methods are briefly described below.

5.3.1 Solar Radiation

The use of sunlight for water purification dates back to at least 2000 B.C. (Patwardhan 1990). Somani and Ingole (2011) pointed out that the germicidal action of sunlight has long been recognized, although the ecological implications of its use and the potential for its practical application to large water distribution systems must be investigated more thoroughly. Several technologies use solar radiation to disinfect water. Solar radiation has been used to inactivate microbes in either dark or opaque containers by relying on heat from sunlight energy (Wegelin et al. 1994; Joyce et al. 1996). Another example is the solar water disinfection (SODIS) system. SODIS uses sunlight (UV spectrum) to penetrate clear plastic containers to achieve disinfection via the combined action of UV radiation, oxidative activity form dissolved oxygen, and heat. Other solar radiation exposure systems also employ combinations of solar effects in UV-penetrable plastic bags; the "solar puddle" and panels (WHO 2011).

The SODIS system is well documented and has been disseminated globally; it is a low-cost water purification method that is fitted to the point-of-use (WHO 2002a, b). Hirtle (2008) performed laboratory studies on SODIS and found it to be highly efficacious for inactivating waterborne pathogens, such as *E. coli*. Pathogenic microorganisms are vulnerable to two different sunlight effects: radiation in the spectrum of UV-A light (wavelength 320–400 nm) and heat (increased water temperature). The UV-A radiation component has a germicidal effect, whereas infrared radiation raises the water temperature and is known to pasteurize water when the water temperature reaches 70–75 °C (Hirtle 2008). The combined use of UV-A radiation and heat produce a synergetic effect that enhances the efficiency of the process. Sunlight and UV light have detrimental effects on many microorganisms and may be a practical method for inactivating viruses, mycoplasma, bacteria, and fungi, particularly those that are airborne. Technologies that are based on solar water disinfection have been field tested in many parts of the world (e.g., Brazil, Colombia, Bolivia, Burkino Faso, Kenya, Tanzania, Ethiopia and Togo, China, Indonesia, and Thailand (WHO 2002a, b).

According to the latest WHO Guidelines for Drinking Water Quality (WHO 2011), under optimal conditions of sunlight, oxygenation, exposure time, temperature, turbidity, and size of water vessel (depth of water), solar disinfection can produce a 3 log reduction in bacterial pathogen load, a 2 log reduction in viral pathogens, and a 2 log reduction in protozoa. The size of the water vessel or depth of water to be disinfected is an important variable during solar disinfection, and study results have indicated that solar water disinfection is effective and feasible for small water quantities (±10 L) of low turbidity as they are easily penetrable (Salih 2003; Mani et al. 2006). The advantage of solar disinfection is that this method also eliminates taste and odor problems and completely destroys microorganisms. However, the successful application of this method depends on the availability of space to set up the apparatus and the existence of uniform solar radiation; this process works best for water of low turbidity. Another limitation with this method is that it does not provide any residual effect, which is a critical necessity for systems having an extensive distribution network or long storage times (SANDEC 2002; LeChevallier and Au 2004).

5.3.2 Membrane Technology

Treating water by using membrane technology has been practical for about 25 years. Membrane processes are used to treat water for achieving multiple purposes, including water clarification, pathogen removal, and removing DBPs or other inorganic and synthetic organic chemicals (Jacangelo et al. 1997; Van der Bruggen et al. 2003). Membrane technologies that are capable of removing microorganisms include microfiltration (MF), ultrafiltration (UF), and reverse osmosis (RO) (Fane et al. 2011). MF membranes have the largest pores, ranging from 0.1 to 10 μm, and the highest permeability so that high water flux is obtained at low pressure. Therefore, MF is an efficient for removing particles that may cause problems in further treatment steps. By contrast, UF membranes have smaller pore sizes (0.002–0.1 μm), lower permeability and require higher pressures to achieve flux than does MF.

MF and UF have been used as alternatives to conventional treatment for removing protozoan cysts from water. It has been generally accepted that MF and UF provides complete removal of all protozoan cysts of concern, as long as the associated system components are intact and operating correctly. Results have indicated that MF and UF membranes provide >4 log to 6 log removal rates of coliform bacteria, *Giardia* spp., and *Cryptosporidium* spp. cysts (Adham and Jacangelo 1994; Jacangelo et al. 1995, 1997; Freeman et al. 1996; Hirata and Hashimoto 1998; Edwards et al. 2001). UF technology is able to remove viruses much more effectively than MF, because of its low cut-off; hence, UF can take the place of the disinfection step (Guo et al. 2010).

Reverse osmosis is normally used for desalination, but it may also potentially be used to disinfect water. Siveka (1966) showed that large numbers of bacteria could be removed by using a reverse osmosis pilot plant. The results obtained show that the feed water coliform concentration was reduced from about 11,000 per mL to less than 3 per 100 mL in the final water. Reverse osmosis utilizes a semipermeable membrane as the separating agent, and pressure as a driving force. However, small fractions of microorganisms do pass through osmosis membranes. Therefore, membranes are not necessarily the ultimate barriers for contaminant removal and require additional disinfection by chlorination or ozonation (LeChevallier and Au 2004; Fane et al. 2011).

Generally, membrane technologies offer an alternative to conventional water disinfection because they produce a high-quality clarified effluent without the need to add chemical reagents, thus avoiding the formation of harmful by-products (Croll 1992; Gomez et al. 2006). Membrane filtering, however, is limited by the high operating costs involved, the fouling potential and their tendency to lose integrity under certain influent conditions (Pianta et al. 2000; Lechevallier and Au 2004). In addition, the singular use of membranes will not guarantee that the water reaching the consumer is safe.

5.3.3 Metal Disinfectants

Metallic ions have been used for water disinfection for some time. Some metal ions, such as that of potassium, silver, and copper are known to possess some disinfection property (Cho et al. 2006). Disinfection by metal ions may result from action at the cell or capsid protein surface or on the nucleic acid of cells or viruses. In addition, metals may alter enzyme structure and function or facilitate hydrolysis or nucleophilic displacement. Potassium permanganate is a strong oxidizing agent and is used to control taste, odors, and biological growth in treatment plants, as well as to remove color, iron, and manganese (Hazen and Sawyer 1992). Potassium permanganate also controls the formation of THMs and other DBPs, by oxidizing precursors and reducing the demand for other disinfectants (Hazen and Sawyer 1992). Despite all these potential uses as an oxidant, potassium permanganate is a poor disinfectant (Bhole 2000).

The use of silver ions to inactivate microorganisms such as, bacteria, viruses, algae, and fungi (in the part per billion ranges) has been widely reported (Chambers et al. 1962; Yamanaka et al. 2005; Jee et al. 2008). The most widely known bactericidal mechanism of the silver ion is its interaction with the thiol groups of the L-cysteine residue of proteins and subsequent inactivation of their enzymatic functions (Russell and Hugo 1994; Liau et al. 1997; McDonnell and Russell 1999). Currently, silver ions are being widely used in health care to control microorganisms, and in water supply systems, since they do not alter the taste, odor, or color of the final water (Cho et al. 2005; Zhang et al. 2005).

Silvestry-Rodriguez and coworkers (2007) found silver to be effective against planktonic bacteria, and silver has been used for water disinfection in Europe (Russell and Hugo 1994; Kim et al. 2004). In addition, Stout and Yu (2003) reported that a combination of silver and copper was effective against *Legionella pneumophila*. Preliminary results of silver electrochemistry experiments suggested that silver has an important role as a safe alternative to chlorine for disinfecting drinking water (Shuval et al. 1995; Batterman et al. 2000, Pedahzur et al. 2000). Currently, two major applications for silver electrochemistry are being investigated that involve municipal drinking water treatment. In both investigations, silver was combined with hydrogen peroxide (HP/Ag) to form a powerful virucidal and bacteriostatic agent. Hydrogen peroxide, in water, liberates nascent oxygen, which possess germicidal properties (Bhagwatula et al. 2000). Although, hydrogen peroxide exhibits high activity, its use as the sole primary or secondary disinfectant in water treatment is not acceptable (Lund 1963; Yoshpe-Purer and Eylan 1968; Toledo et al. 1973).

Batterman et al. (2000) studied the potential for applying an HP/Ag formula as an additional disinfection treatment processes. Results show that an engineered method that employs the HP/Ag system can be effective when used after chlorination. This HP/Ag method dramatically decreased the toxicity of chlorine treatment by quenching the formation of DBPs. The HP/Ag water disinfection method may work in large scale water systems as contact times are sufficiently long. Pedahzur et al. (2000) similarly concluded that the HP/Ag method may be used for distribution systems where the time between treatment and point of use is over 900 min. Zhou and coworkers (1997) found that silver does not react with most organics in water to produce toxic by-products. However, the extent to which silver alone inactivates microbes in water is limited, because microorganisms may develop silver resistance over time. Therefore, more research is needed to warrant endorsement of this method for routine disinfection of drinking water. In addition, compared to conventional disinfection methods, silver disinfection is not commercially economical.

5.3.4 Ultrasonic Disinfection

Sonication may disintegrate cells by cavitation; hence, ultrasonic waves have been used to inactivate microorganisms as means of disinfection (Wong 2002). Unfortunately, ultrasound technology is substantially more expensive than are alternate methods of disinfection (Sassi et al. 2005).

5.3.5 Halogens

Halogens, other than chlorine, like iodine and bromine also exhibit a disinfection property (CPHEEO 1999). The effects of iodine as a disinfectant are similar to those of chlorine, although its commercial use is mainly limited by its cost, effects on the thyroid gland, potential toxicity, and allergenicity (Backer and Hollowell 2000; Goodyer and Behrens 2000). The main advantage of using iodine over chlorine is that it has a more acceptable taste. Iodine, like chlorine, is employed for emergency use and use under certain field conditions (Gerba et al. 1997). Bromine, like chlorine, is a disinfectant. Once dissolved in water Br produces hypobromous acid (HOBr) (WHO 2003a, b). The disinfecting power of HOBr is high, but slightly less than that of hypochlorous acid. However, Br is highly reactive with ammonia and other amines, and this may seriously limit its effectiveness under typical water treatment conditions. Bromine also forms trihalomethanes, and if fulvic acids and ammonia are present in raw water, bromoform, a potential carcinogen, is produced (WHO 2003a, b).

5.3.6 Applications of Nanotechnology in Water Disinfection

Nanotechnology is an emerging technology that has significant potential for application in many areas, including drinking water disinfection. Nanotechnology has been used to revolutionize and improve the century-old conventional disinfection methods (USEPA 2007; Shannon et al. 2008). The major property that makes nanomaterials attractive is that they are of extremely small size (1–100 nm), which provides them with a higher reactive surface area per unit mass. Nanomaterials also show unique and enhanced physical, chemical, and biological properties, compared to their macro-scaled counterparts (Heidarpour et al. 2011). In recent years, nanomaterials have been introduced to the water industry that may be promising for disinfection purposes. Natural and engineered nanomaterials that have strong antimicrobial properties include chitosan (Qi et al. 2004), metallic nanoparticles (e.g., silver ion) (Morones et al. 2005), photocatalytic TiO_2 (Cho et al. 2005; Wei et al. 1994), nanofilters (Van der Bruggen and Van der Casteele 2003; Srivastava et al. 2004), aqueous fullerene nanoparticles (Lyon et al. 2006; Badireddy et al. 2007), and carbon nanotubes (Savage and Diallo 2005; Kang et al. 2007).

De Gusseme et al. (2010) stated that silver-based nanotechnological agents have received considerable attention for water disinfection. They have shown in laboratory studies that the addition of 5.4 mg/L of zerovalent biogenic silver nanoparticles resulted in a 4 log decrease of the targeted phage after 1 h of contact. The antiviral properties of biogenic Ag^0 were also demonstrated against the murine norovirus1 (MNV-1), a model organism for human noroviruses. In another study, Savage and Diallo (2005) demonstrated that nanosorbents such as carbon nanotubes (CNTs), polymeric materials (e.g., dendrimers), and zeolites have exceptional adsorption properties and can be applied to remove biological impurities. CNTs, in particular, received attention for exceptional water treatment capabilities and proved to be effective against a range of microorganisms, including bacteria. Many researchers have attributed the success of CNTs as an adsorbent for removing pathogens mainly to its unique physical, cytotoxic, and surface functionalizing properties (Akasaka and Watari 2009; Deng et al. 2008; Srivastava et al. 2004; Upadhyayula et al. 2008a,b; Mostafavia et al. 2009), and to its toxicity to cyanobacteria (Albuquerque et al. 2008; Yan et al. 2004, 2006).

Nanostructured membrane processes such as NF are also emerging as key components of advanced water purification (US Bureau of Reclamation 2003). Van der Bruggen and Van der Casteele (2003) reported that nanofilters provide novel opportunities for water purification and are an efficient and cost effective method for removing cations, natural organic matter, biological contaminants, organic pollutants, nitrates, and arsenic from groundwater and surface water. Peltier et al. (2003) demonstrated the potential for nanofiltration as means to reduce large quantities of both organic and biological contaminants (e.g., bacteria and viruses) in a large water distribution system.

From these studies, it is evident that nanomaterials have several key physicochemical properties that allow them to disinfect water. Nanomaterials also present exceptional opportunities for development of more efficient, cost effective, and reliable water disinfection technologies as a result of their large surface reactive areas and their geometric, electronic, and catalytic properties. Savage and Diallo (2005) studied the current and future strategies for using nanomaterials as alternative disinfecting agents. This author's work emphasized the development of chlorine-free biocides through chemical group functions that selectively target key biochemical constituents of waterborne bacteria and viruses. Currently, the Desalination and Water Purification Roadmap prepared by the US Bureau of Reclamation and Sandia National Laboratories is conducting research to develop smart membranes that have biofilm resistant surfaces and embedded sensors/actuators that automatically adjust membrane performance and selectivity (US Bureau of Reclamation 2003). The foregoing shows that, in the near future, nanomaterials are positioned to become critical components for providing safe drinking water.

5.4   Formation and Control of Disinfection By-Products

All chemical disinfectants produce inorganic or organic DBPs that may be of health concern. Factors affecting DBP formation include the nature of the disinfectant, the dose, mixing efficacy, presence and composition of natural organic matter, concentration, pH, time that the disinfectant is in contact with the organic matter, water temperature, bromide ion concentration, and to a lesser extent the concentrations of iodide, and nitrite, or organic nitrogen (applicable to chloropicrin formation) (Kawamura 2000). These factors, in turn, depend on both the water quality and the treatment process; hence, variations in either water quality or treatment will create changes in DBP levels. Each of these factors is briefly discussed below.

- *The amount of natural organic matter present.* If there is no organic precursor in the water, there will be no THM formation. The concentration and type of precursors directly influence THM formation. Chemical disinfectants oxidize naturally occurring organic and inorganic precursors in drinking water to produce DBPs. The primary precursor is natural organic matter (NOM), which is generally measured as total organic carbon (TOC), and is composed of approximately 50% carbon, 35% oxygen, 5% hydrogen, 3% nitrogen, and lower amounts of phosphorous, sulfur, and trace metals (Westerhoff 2006). Bromide is an important precursor, because bromide is easily oxidized to aqueous bromine (HOBr/OBr$^-$), a mild disinfectant that can react to form bromine-substituted DBPs such as bromoform.
- *The disinfectant used.* Chlorine has a more profound effect on THM formation than does chloramines (Table 11). The disinfectant dose also affects THM formation, because the amount of chlorine residual in the system has a direct impact on DBP formation.
- *Water pH.* The higher the pH, the faster the reaction rate, and the higher the THM levels will be. This phenomenon is believed to result from pH-induced changes in the functional groups of the precursor molecules (Kawamura 2000).

Table 11 Disinfection by-products present in disinfected waters (WHO 2011)

Disinfectant	Organohalogens	Inorganics	Non-organohalogens
Chlorine	Trihalomethanes Haloacetic acids Haloacetonitriles Chloral hydrate Chloropicrin Chlorophenols *N*-chloramines Halofuranones Bromohydrins	Chlorate (mostly from hypochlorite use)	Aldehydes Cyanoalkanoic acids Alkanoic acids Benzene Carboxylic acids *N*-nitrosodimethylamine
Chloramine	Haloacetonitriles Cyanogen chloride Organic chloramines Chloramino acids Chloral hydrate Haloketones	Nitrate Nitrite Chlorate Hydrazine	Aldehydes Ketones *N*-Nitrosodimethylamine
Chlorine dioxide		Chlorite chlorate	Unknown
Ozone	Bromoform Monobromoacetic acid Dibromoacetic acid Dibromoacetone Cyanogen bromide	Chlorate Iodate Bromate Hydrogen peroxide Hypobromous acid Epoxides Ozonates	Aldehydes Ketoacids Ketones Carboxylic acids
UV	Unknown	Unknown	Unknown

- *Water temperature.* Water temperature greatly affects the rate of THM formation. The higher water temperature induces a faster reaction rate, which, in turn, leads to more THM (Kawamura 2000).
- *The time available for reaction.* The THM level is a function of chlorine contact time. The longer the contact time, the higher the level of THM (Kawamura 2000).

Although >500 different DBPs have been reported in the literature, more research is needed to uncover and identify new DBPs. Currently, the DBPs that have been quantified in drinking water account for only about 40% of the total organic halide (TOX) concentration measured in chlorinated drinking water, and about 40% of the assimilable organic carbon (AOC) concentration measured in ozonated drinking water (Richardson et al. 2002). In addition, TOX and AOC represent only a portion of the types of DBPs that are formed.

Some of the known principal organohalogen DBPs formed during chlorination include the trihalomethanes, haloacetic acids, haloacetonitriles, chloral hydrate, chloropicrin, chlorophenols, *N*-chloramines, halofuranones, and bromohydrins. Although chloramination of drinking water produces lower THM concentrations (if chloramines are formed post-ammoniation) than does chlorine, using it may produce other DBPs, such as HAAs, chloral hydrate, hydrazine, cyanogen compounds, nitrate, nitrite, organic chloramines, and 1,1-dichloropropanone (1,1-DCPN) (Dlyamandoglu and Selleck 1992; Kirmeyer et al. 1993, 1995). The presence of even small quantities of organic nitrogen in chloraminated drinking water results in the possible formation of organic chloramines, which are much weaker disinfectants than are the inorganic form (Isaac and Morris 1983; Bercz and Bawa 1986).

Ammonia is released to the distribution system, primarily from complex auto-decomposition reactions of chloramines, and produces by-products such as nitrite and other nitrogenous compounds (Woolschlager et al. 2001). Nitrate and nitrite formation in the presence from nitrifying bacteria is among the most challenging problems in chloraminated drinking water supplies (see Sect. 5.6). For health reasons, it is important that nitrite and nitrate levels in drinking water not exceed the current regulatory requirements. One other possible health risk associated with ammonia being present in water is the formation of carcinogenic nitrosamines from organic amines (e.g., nitrosodimethylamine (NDMA)) (Jang 2009).

One disinfectant that is an alternative to chlorine, when trying to avoid disinfectant residuals, is ozone. Ozone disinfects by oxidation, and effectively kills microorganisms, and does not usually produce THMs or other chlorinated DBPs (Caughran et al. 1999). The major drawbacks of using ozone are that it oxidizes bromide to hypobromite and bromate, and hypochlorite to chlorate (Glaze et al. 1993; Siddiqui et al. 1995; Siddiqui et al. 1996). Such brominated compounds are of health concern because they are known to be more toxic than the corresponding chlorinated by-products (Richardson 2003).

Chlorine dioxide is an alternative to ozone and chlorine for disinfection of water. Unlike chlorine, chlorine dioxide does not provide a disinfectant residual. The health concerns for using chlorine dioxide is that it leaves residual concentrations of chlorine dioxide itself and the by-products chlorite and chlorate, although these can be controlled by limiting the dose use at the treatment plant (WHO 2000).

UV irradiation has long been considered as the most plausible alternative for disinfection, when avoiding formation of DBPs is sought. Unfortunately, UV light can result in the conversion of nitrate to nitrite. In addition, UV disinfection forms low levels of aldehydes and carboxylic acids (Malley et al. 1995; Peldszus et al. 2004; Thomson et al. 2002; Liu et al. 2002). Because UV radiation only works in real time, an additional disinfectant would be subsequently needed to assure the presence of residual disinfectant ion in the treatment distribution system. We should also point out that disinfection efficiency should never be compromised just to meet guideline levels for DBPs, including chlorination by-products, or in trying to reduce concentrations of these substances.

Although guideline values have been established for many DBPs, data from drinking water supplies indicate that the THMs and HAAs that are detected in water are good indicators for the levels at which the majority of other by-products will exist. To meet DBP regulations a change disinfectant may be needed, although this is not always possible. For example, to reduce THM formation a change from chlorine to monochloramine may provide residual disinfection within a distribution system. Although monochloramine provides a more stable residual effect, it is a less powerful disinfectant and should not be used as the primary disinfectant. Other strategies to reduce DBP levels include controlling it at the source, removing precursors, removing DBPs by air stripping, activated carbon filtration, or employing UV light and advanced oxidation treatment approaches (WHO 2000).

5.5   Biofilm Formation and Control

The basis for most water distribution system problems are microorganisms; such microbe-linked problems include nitrification, biofilm growth, microbially induced corrosion, and the persistence of pathogens (Regan et al. 2003; Beech and Sunner 2004; Camper 2004, Emtiazi et al. 2004). Biofilms are believed to be the primary source of microorganisms in distribution systems, and their ability to metabolize recalcitrant organic compounds, and their increasing resistance to chlorine and other biocides helps them to withstand the conditions that exist in water distribution systems (Camper 2004, Emtiazi et al. 2004; Tachikawa et al. 2005). The amount of nutrient present is the most important factor that affects biofilm growth in distribution systems. Carbon, nitrogen, and phosphorus are essential for the growth of heterotrophic bacteria in distribution systems. Organic carbon is the limiting factor for biofilm microbial growth. Therefore, controlling the type and concentration of organic carbon in the distribution system greatly affects the growth potential for biofilms (Emtiazi et al. 2004).

Biofilm growth is associated with warmer temperatures and may also be linked to the seasons. The temperature relationship is complicated by the fact that most fecal bacteria survive disinfection longer at lower temperatures. In some systems, rainfall events have also been associated with increased biofilm growth, due to higher concentrations of nutrients, turbidity, and bacteria in the source water that leads to treatment breakthrough. Biofilm development is also linked to the types of materials that exist in distribution systems (Emtiazi et al. 2004). Biofilms grow more rapidly, more densely, and more diversely on iron pipe, particularly older piping, compared to PVC (polyvinylchloride) pipes. Other components that support biofilm growth include the materials that comprise valves, gaskets, washers, pump lubricants, and pipe coatings (Emtiazi et al. 2004).

Biofilms influence the taste and odor of water, and when they grow on ferrous metal surfaces may corrode pipes and release iron particles into the water (Ridgway et al. 1981; Camper et al. 1999). Biofilms also harbor pathogens (Falkinham et al. 2001; September et al. 2007). Therefore, during turbulent water flow (e.g., flushing), pathogens may be dislodged from the biofilm and enter the bulk water flow, placing consumers at risk (September et al. 2007). Surveys of water distribution networks that utilize polymerase chain reaction (PCR) techniques and Southern blot hybridization have shown the presence of bacterial pathogens, including *Legionella* spp., *Cryptosporidium* spp., *Helicobacter* spp., *Mycobacterium* spp. (*M. avium* and *M. intracellulare*), and *Aeromonas* spp., as well as viral pathogens, such as enteroviruses and adenoviruses (Falkinham et al. 2001; Park et al. 2001; Nichols et al. 2003; Schwartz et al. 2003; Emtiazi et al. 2004; Sen and Rodgers 2004; Skraber et al. 2005). It is for this reason that water regulations specify the frequency for routine monitoring of pathogens.

In contrast to the foregoing, Mains (2008) argued that although primary pathogens have been detected in biofilms, there is little conclusive evidence that links biofilms to waterborne disease outbreaks. Notwithstanding, these authors note that biofilms are of great concern to people with weakened immune systems, such as AIDS patients, diabetics, organ transplant recipients, many cancer patients, and the elderly and young children. Opportunistic pathogens that have been identified in biofilms include *Pseudomonas aeruginosa, Legionella pneumophila,* and the Mycobacterium avium complex (MAC). Apart from pathogenic microorganisms, Tokajian and coworkers (2005) revealed that several novel bacterial strains are known to exist in chlorinated waters; among these that have been isolated are gram-positive bacteria and alpha-, beta-, and gamma-proteobacteria. Williams et al. (2004) found alpha-proteobacteria to be dominant in both chloraminated and chlorinated water, whereas beta-proteobacteria were more abundant in chloraminated water than in chlorinated water. Similarly, Emtiazi et al. (2004) used 16S rRNA gene-directed PCR and denaturing gradient gel electrophoresis (DGGE) to show that beta-proteobacteria were abundant in biofilms of non-chlorinated distribution systems. In essence, these results indicate that disinfection strategies do affect microbial community diversity in treated water distribution systems.

Nitrifying organisms, belonging primarily to the alpha-, beta-, and gamma-proteobacteria have been the subject of distribution network investigations, because they contribute to nitrification. *Nitrosomonas* spp., members of the beta-proteobacteria, have been identified as the dominant ammonia-oxidizing bacteria (AOB), while the *Nitrosospira* spp. were the dominant nitrite-oxidizing bacteria (NOB) in bulk water and biofilms from pilot- and full-scale distribution systems. *Nitrobacter* spp. also oxidizes nitrite into nitrate and belongs to the alpha-proteobacteria group; this genus was also detected in biofilms of chloraminated drinking water (Regan et al. 2002, 2003; Lipponen et al. 2004). Oldenburg et al. (2002) suggested that the widespread presence of nitrifiers in bulk water distribution systems resulted from the long contact times nitrifying bacteria require to succumb to typical monochloramine doses. Pintar and Slawson (2003) suggested that nitrification processes in the distribution system can be effectively thwarted only through very high or very low chloramine levels, because of the disinfection action at high levels and the scarcity of ammonia at low levels for AOB.

5.6 *Nitrification*

In chloraminated water distribution systems, nitrification may occur from the presence of nitrifying bacteria (Regan et al. 2003; Lipponen et al. 2004; van der Wielen et al. 2009). Nitrification is primarily carried out by two groups of bacteria via a two-step process. The first step involves AOB, which oxidize ammonia to nitrite. In the second step, nitrite is further converted to nitrate by NOB. Recently, Zhang et al. (2008, 2009b) suggested that there is a possible third step, in which the nitrate is "recycled" to ammonia via reactions with corrosion products. More particularly, free ammonia will typically be available to nitrifiers in the distribution system if the chlorine to ammonia ratio is not sufficiently controlled during monochloramine formation (Odell et al. 1996). The presence of ammonia in the distribution system is also a result of the auto-degradation of monochloramine, a process which occurs within days after monochloramine is applied. Monochloramine degrades in a water system by three mechanisms ((1)–(3) below). These mechanisms are (1) auto-decomposition reactions, (2) reactions between organic matters or microbial cells and their metabolic products, and (3) reactions with nitrite (Vikesland et al. 2001; Woolschlager et al. 2001; Yang et al. 2008).

Auto-decomposition

$$3NH_2Cl \rightarrow N_2 + NH_3 + 3Cl^- + 3H^+ \tag{1}$$

Oxidation of organic matter

$$0.1C_5H_7O_2N + NH_2Cl + 0.9H_2O \rightarrow 0.4CO_2 + 0.1HCO_3^- + 1.1NH_4^- + Cl^- \tag{2}$$

Reduction by nitrite

$$NH_2Cl + NO_2^- + H_2O \rightarrow NH_3 + NO_3^- + HCl \tag{3}$$

The auto-decay of chloramines (1) is a principal cause of a high nitrification rate in chloraminated distribution systems. Nitrification is affected by chloramine degradation reactions and the presence of excess ammonia and nitrifying microorganisms. In addition, other factors that affect nitrification include pH, temperature, retention times, and disinfectant residuals. Most chloraminated water distribution systems are more susceptible to nitrification during the summer months, when temperatures reach 15 °C or higher. Higher temperatures increase the growth rate of nitrifying bacteria and thus the risk of nitrification (Antoniou et al. 1990; Rittmann and Snoeyink 1984; Pintar and Slawson 2003; Yang et al. 2008). Additionally, Vikesland et al. (2001) revealed that higher temperatures increase the chemical decay rate of monochloramine. Nitrification can also occur in colder waters at pH ranging from 6.5 to 10.0 (Wilczak et al. 1996). The effect of pH on nitrification is profound and complicated; it occurs by several mechanisms, which at times may act in opposition. Zhang et al. (2009b) described some of the diverse ways in which pH affects nitrification, e.g., changing the balance of ammonia and ammonium (free ammonia is thought to be the true substrate), causing inorganic carbon loss from CO_2 stripping at low pH values that affects chloramine decay rate, affecting the growth of nitrifiers, and changing the rate of nitrifiers disinfection by chloramines (Zhang et al. 2009b).

Of more relevance is the effect of pH on monochloramine decay rate. According to Vikesland et al. (2001), at lower aqueous pH, the rate of monochloramine auto-decomposition was more rapid. In addition, higher carbonate concentrations led to more rapid monochloramine decay at a given pH, which led to the conclusion that monochloramine auto-decomposition is increased under acidic conditions. However, the effect of pH on the disinfection efficiency of monochloramine appears to act in opposition to the impact of pH on the stability of the monochloramine disinfectant residual. Increased pH results in the formation of other chloramines (di- and tri-chloramine), and chloramines have lower disinfection capacity than monochloramine.

The initial conversion of ammonia to nitrite (incomplete nitrification) is the most critical step in nitrification, due to the role of this step in the degradation of the disinfectant residual. Incomplete nitrification produces several adverse effects on water quality. These include decreased DO concentration, increased numbers of bacteria, and increased nitrite and nitrate concentrations (Federation of Canadian Municipalities and National Research Council 2003). According to the US Environmental Protection Agency (USEPA 2010), the presence of nitrite in distribution systems causes many technical problems for disinfection. Nitrite reacts with the disinfection residuals (3) to increase the microbial populations and organic matter. In turn, the organic matter can significantly increase the chlorine demand of the water, resulting in a rapid loss of disinfectant residual. Similarly, Pintar et al. (2005) observed a partial loss of the total chlorine residual in a full-scale distribution system that had high nitrite levels, suggesting that decreased total chlorine residual can serve as an early warning of a developing nitrification episode.

During the nitrification process, nitrite is converted to nitrate by NOB; however, bacterial mediation of this step is less critical because a chemical-oxidation pathway also exists. This pathway involves chemical oxidation of nitrate by chloramine (Vikesland et al. 2001; Yang et al. 2008). The conditions under which nitrite is converted are less oxidative, which in turn promotes the growth of biofilm-active bacteria (i.e., AOB and NOB). These and other heterotrophic bacteria that proliferate under lower oxidative conditions complicate maintaining the chloramine residual in water distribution systems (Skadsen 1993). Therefore, most drinking water regulations recommend that water utilities monitor heterotrophic plate counts, particularly in drinking water distribution systems that are chloraminated. Pintar et al. (2005), in contrast, did not observe a correlation between HPCs and the onset of nitrification in a chloraminated distribution system.

Therefore, it is apparent that there are several factors that can lead to high HPCs besides nitrification, and isolated HPCs cannot be used as a nitrification indicator. Zhang et al. (2009b) noted that pH is one factor that can indicate the occurrence of nitrification. As more ammonia is oxidized, the alkalinity of water is consumed, resulting in a drop in pH. In turn, low pH may affect corrosion in the distribution system. Apart from the pH drop, nitrification also affects the taste and odor qualities of water, and often produces increased complaints from consumers. The foregoing notwithstanding, nitrification is not likely to affect public health, but rather may lead to operational or regulatory compliance challenges. An intrinsic consequence of nitrification is to increase nitrite and nitrate levels. Wilczak et al. (1996) reported that nitrite and nitrate are probably the most frequently recommended indicators of nitrification. This is because there is more background variability in their levels between water leaving the plant and water in the distribution system.

From a health perspective, the presence of elevated levels of nitrites and nitrates in final drinking water is of less concern than is the loss of chloramines residual.

Increased nitrates and nitrites are more important for certain subpopulations, such as infants (less than 6 months). Intake by infants of nitrite and nitrate have been shown to induce methaemoglobinemia (blue baby syndrome), an acute response to nitrite that results in a blockage of oxygen transport (Bouchard et al. 1992). Moreover, the low pH conditions in the stomach reduce nitrate to nitrite, which then reacts with secondary amines to produce nitrosamines, which are known carcinogens (Bouchard et al. 1992; De Roos et al. 2003). Therefore, both nitrite and nitrate are regulated as drinking water contaminants; the WHO Guidelines for Drinking Water Quality recommends a maximum acceptable limit of 0.9 and 11 mg/L, for nitrite and nitrate, respectively. The USEPA stipulates an MCL level for nitrite as 1 mg/L and nitrate as 10 mg/L. According to the European Council Directive 98/83/EC, the maximum acceptable drinking water concentration of both ammonium and nitrite is 0.50 mg/L and that for nitrate is 50 mg/L. Nitrification poses a risk to public health. Therefore, controlling nitrification is now a major issue and is becoming increasingly important as an increasing number of water utilities adopt chloramination for water disinfection.

The most common method for controlling nitrification is to achieve breakpoint chlorination. During breakpoint chlorination all ammonia present in the system is used-up, resulting in free chlorine residual. However, according to Odell et al. (1996), the main disadvantage of breakpoint chlorination is that it may increase the number of microorganisms in the final drinking water as a result of increased biofilm detachment. Therefore, when dense biofilm layers are present, breakpoint chlorination should be used as a last option. Maintaining a higher chloramine residual at strategic points within the water distribution system can also be effective in halting nitrification (Skadsen 1993; Odell et al. 1996; Pintar and Slawson 2003). McVay (2009) recently recommended that water utilities target a chloramine residual of ≥2.5 mg/L in stagnant areas of the distribution system, because it significantly hinders nitrification. While raising the chloramines levels to >2.5 mg/L is the best way to impede nitrification, water utilities must ensure they still comply with disinfection target regulations. For example, the USEPA drinking water rules prohibit chloramine concentrations in a distribution system from exceeding 4.0 mg/L. In large distribution systems, or during seasons of low water demand, a 2.5-mg/L residual can be maintained by using additional chloramine dosing facilities located closer to the identified problem areas. Alternatively, maintaining a combined chlorine residual of at least 1 mg/L also minimizes the susceptibility of distribution systems to nitrification.

Another method used successfully by water utilities for maintaining chloramine residuals is to maintain an alkaline pH. A pH increase is effective for maintaining chloramine residuals because it slows ammonia release from chloramine degradation. However, there are few systems that have the ability to adjust pH without adding an additional treatment process. Therefore, this method is not easily applied. As an alternative, other operational activities, which include reducing the residence time, eliminating water stagnation water within the distribution system, and maintaining higher chlorine to ammonia ratios (up to 5:1), have potential. In particular, flushing can be an effective method for eliminating water stagnation, while also moving fresh chloramine and improving the oxidation in the water distribution system. However, this approach should not be attempted without developing a flushing plan that identifies the amount of water needed for flushing and the duration over which the pipe segment must be flushed to achieve the chloramine residual sought. A flushing program for a distribution system is labor intensive, although automated flushing valves can be used (USEPA 2010).

5.7   Disinfectant Residual Loss

Because microbial contamination of water in distribution systems is anticipated, water utilities normally aim to achieve a detectable disinfectant residual so as to minimize the potential for waterborne disease and biofilm growth. However, if the disinfection process is not properly maintained, residual disinfectant loss may occur. Such loss is possibly one of the most serious aspects of drinking water quality deterioration in a distribution network. The loss of disinfectant residual occurs by many different pathways that are poorly understood. However, when residual loss occurs, it may lead to bacterial regrowth as well as other problems such as taste and odor problems.

Results of several studies suggest that the temporal and spatial loss of free-chlorine residual is caused by chemical reactions between the chlorine and water constituents or with both the biofilm and tubercles formed on pipe walls or pipe-wall material (Zhang et al. 1992; Clark et al. 1994; DiGiano and Zhang 2005). These researchers believe the water constituents that react with chlorine and lead to its loss include deposits, corrosion by-products, microorganisms, organic impurities, ammonia compounds, and metallic compounds, such as iron (ferrous ions) and manganese. Previous studies have shown that iron corrosion scales generally contain reduced iron, which can react with oxidative disinfectants, resulting in their loss (Sarin et al. 2001, 2004). Williams (1953) revealed that the loss of both chlorine and chloramine residuals in the Brantford system was related to an increase in ferrous iron caused by corrosion in the distribution system pipes.

Similarly, Powell (2000) concluded that water distribution systems fitted with cast-iron pipe material exhibited higher chlorine residual losses (typically 10–100 times higher) than did PVC pipes. The loss of chlorine dioxide residual (chlorite) occurred in cast-iron pipe loops and in full-scale drinking water distribution systems containing cast-iron pipes (Baribeau et al. 2002; Eisnor and Gagnon 2004). Zhang et al. (2008) identified both geothite (a-FeOOH) and magnetite (Fe_3O_4) as the main components of iron corrosion scale, whereas cuprite (Cu_2O) was identified as the major component of copper corrosion scale. From this study, it was concluded that the presence of both iron and copper oxides produced a loss of chlorine dioxide residual in the system. However, the loss of ClO_2 in the corroded copper pipe was lower than that in the iron pipe.

It was discovered that hydraulic/operating conditions, including flow velocities and water residence time, may affect the microbiological quality of the water. This occurs when the residual disinfectant is consumed to produce increased microbial growth. Chlorine residual loss experiments were conducted, using 70–135 year old unlined cast-iron pipes, under varying flow conditions; results showed that higher flow rates resulted in greater residual chlorine loss in the pipe (Grayman et al. 2002; Doshi et al. 2003). Moreover, certain hydraulic conditions also favor the deposition and accumulation of sediments, and consequently enhance microbial growth and effect residual loss by protecting microbes from disinfectants.

Maul et al. (1985a) showed that both free and total chlorine residuals in water rapidly decreased in a distribution system as residence time increased. Furthermore, the rate of free chlorine residual loss increased with increasing water temperature. Similarly, a pilot study by Arevalo (2007) showed that the rate of free chlorine and chloramine decay was highly affected by the pipe material; the decay was faster in unlined metallic pipes and slower in the synthetic (PVC) and lined pipes. Additionally, the rate of disinfectant residual loss was increased with increasing temperature and organics content in the water, irrespective of pipe material. The causal factors that influence disinfectant residual loss in distribution network mains are quite complex and intertwined. Therefore, maintaining a disinfectant residual as means to achieving effective water distribution network disinfection is a difficult task. This is because one factor is the disinfectant process itself, and others are the constituents present in the treated water, the season, and, among others, the thousands of kilometers of pipes of different ages and materials that make up the water distribution network.

6   Risk Management in Drinking Water Disinfection

6.1   Basic Concepts in Drinking Water Safety

The main goal of disinfection is to eliminate pathogenic organisms that are responsible for waterborne disease. Drinking water disinfection typically includes multibarrier water treatment processes such as settlement, coagulation, and filtration, as well as a final disinfection by a process such as chlorination. However, hardly any treatment process can be trusted to always perform the same, because of the existence of variables like turbidity, temperature, microbial load, and pH. Extreme cases occur, where the nominal (designed) performance of a treatment process will break down. Real patterns of variation have been demonstrated to exist that produce or influence the distribution of risk in exposed population (Teunis et al. 1997; Teunis and Havelaar 1999).

There have been documented examples in which pathogens were not sufficiently eliminated during water treatment or in which the treatment system failed. Such incidents, though uncommon, have created outbreaks, such as the following: the Milwaukee incident of 1993, in which *Cryptosporidium* contaminated the public water supply and caused >100 deaths and an estimated 403,000 illnesses (Mackenzie et al. 1994); the Walkerton incident of 2000, in which >2,300 individuals were diagnosed with gastroenteritis, 65 were hospitalized, 27 developed hemolytic uremic syndrome (HUS; a serious and potentially fatal kidney ailment), and 7 died from drinking water contaminated mainly by *E. coli* 0157:H7 and *Campylobacter jejuni* (Hrudey et al. 2003).

Such events prompted implementation of microbial risk management systems that do not rely on water quality monitoring results for ensuring drinking water safety. Percival et al. (2000) stated that risk management in drinking water is important, because it helps to predict the burden of waterborne disease in communities faced with outbreak and non-outbreak conditions. Additionally, microbial risk assessments help policy makers and water utilities to establish microbial standards for drinking water supplies to minimize health effects to exposed populations. Performing risk assessments also help to identify the most cost-effective treatment option to reduce any microbial risk, while balancing infection with chemical risks from DBPs. In addition, risk communication (proactive and reactive) plays an essential role in achieving the goal of having safe drinking water and is an essential element of risk management itself. Sahota and Pandove (2010) suggested that there is also need to improve on current risk-assessment modeling protocols, and dissemination of information on water testing so as to coordinate surveillance systems for early detection, tracking, and evaluation of emerging waterborne pathogens.

Teunis and coworkers (2009) proposed that mathematic models be used to accurately estimate the variation in microbial log reduction (e.g., probability distributions). Moreover, they suggested that system design and location, maintenance practices, and employee awareness be critical elements of a successful microbial risk reduction program. Another approach for minimizing the drinking water risk involves developing tools to aid the management of disinfection in water distribution systems; having such tools has been a high priority for many water utilities around the world. Several instruments have recently been developed for safeguarding the microbial quality of drinking water, and using them has proved successful in various parts of the world. A few of these are briefly described below.

6.2   HACCP and Water Safety Plans

The Hazard Analysis Critical Control Point (HACCP) approach was developed for the food industry to improve safe food; however, this framework has since been adopted by the water supply industry (Deere et al. 2001; Dewettinck et al. 2001; Howard et al. 2003; Westrell et al. 2004). The main principles for HACCP are to (1) identify hazards and preventive measures, (2) identify critical control points, (3) establish critical limits, (4) identify monitoring procedures, (5) establish corrective action procedures, (6) validate and verify HACCP plans, and (7) establish documentation and recordkeeping practices. The risk-based Water Safety Plans (WSPs) recommended in the third edition of the WHO Guidelines for Drinking Water Quality, employ many of these principles and concepts that

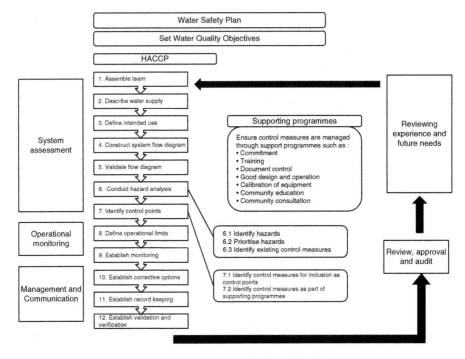

Fig. 1 Steps in the development of a water safety plan (WHO 2005)

have been taken from other risk management approaches, in particular from HACCP and the multi-barrier approach.

A Water Safety Plan (WSP) essentially consists of three components: (1) system assessment, (2) operational monitoring and management plans, and (3) documentation and communication. These components can be broken down into a series of steps as shown in Fig. 1. The WSP principles provide a framework for assessing and managing microbial risks in any circumstance. The WSP framework also facilitates an increased awareness and understanding of risk issues for providing safe drinking water. The Bonn Charter for Safe Drinking Water (developed over several years by drinking water regulators, drinking water service providers, and drinking water professionals from around the world through the International Water Association) also speaks to the need for integrated water safety plans, based on a risk assessment and risk management process from catchment (source) to tap. The benefits of developing and implementing a WSP for water supplies include the systematic and detailed assessment and prioritization of hazards and the operational monitoring of barriers or control measures. In addition, it provides for an organized and structured system to minimize the chance of failure through oversight or lapse of management (WHO 2005).

6.3 Quantitative Microbial Risk Assessment

The main purpose of the QMRA (quantitative microbial risk assessment) approach is to calculate the risk of disease in a population from what is known or can be inferred from the concentration of specific pathogens in the water supply and the infectivity of these pathogens in humans (Hunter et al. 2003). The ILSI Framework for Microbial Risk Assessment (ILSI 1996) identifies three steps for conducting a QMRA, namely, (1) problem formulation—a planning step to identify goals, regulatory and policy context, and develop conceptual models, (2) analysis—a step to technically evaluate exposure and health effects data, and lastly (3) Risk Characterization—formulating the problem and analyzing the steps to estimate and describe risks. The QMRA approach is now utilized worldwide by policy makers and industry to project expected health risks for existing or planned catchments, treatment facilities, and distribution systems (WHO 2004). Microbial risk assessments generate more robust data on microbial behavior/survival/transport/persistence/virulence/and dose–responses for a broader range of environments (Howard et al. 2006).

Generally, QMRA studies are useful for monitoring pathogen concentrations in raw water, and modeling removal- or inactivation treatment to estimate concentrations in the drinking water (Teunis et al. 1997; Haas and Trussell 1998; Teunis and Havelaar 1999; Westrell et al. 2004). QMRAs that are performed for waterborne pathogens often identify water treatment as the critical stage, with respect to both nominal risk and uncertainty (Teunis et al. 1997). For example, in the Netherlands, the QMRA approach has been successfully applied to estimate infection risks from contamination resulting from insufficient treatment of source water (Dechesne et al. 2006; Smeets et al. 2006). To perform a reliable QMRA for a certain pathogen in the drinking water supply, and for a given population, information on pathogen concentrations in the source water, removal or inactivation efficiency of the treatment process, consumption of drinking water and any special characteristics in a population is required (Teunis et al. 2010).

To date, QMRA has been successfully used to characterize the risks posed by pathogens of public health concern that include viruses (e.g., rotavirus, poliovirus, echovirus, adenovirus, hepatitis A, and Coxsackie virus), parasites (e.g., *Cryptosporidium* and *Giardia*), and bacteria (*Salmonella, Shigella, E. coli* O157, *Vibrio cholera, Legionella pneumophila*, and *Campylobacter jejuni*) (Gerba and Haas 1988; Regli et al. 1991; Rose et al. 1991; Teunis et al. 1994; ILSI 1996; Medema et al. 2003b WHO 2009).

6.4 Disinfection Management Tools

In Australia, to safeguard public health, there is a high priority to develop tools to better manage disinfection of water in distribution systems. The need for disinfection management tools has arisen from the increasingly rigorous demands placed on water utility managers to better manage their systems, as highlighted by the

Framework for Management of Drinking Water Quality (WQMF) incorporated within the 2004 Australian Drinking Water Guidelines (ADWG). Water utilities are challenged to balance compliance for both disinfectant residuals levels and microbiological parameters at the customer tap, while minimizing formation of DBPs and presence of taste or odor compounds. The Cooperative Research Centre for Water Quality and Treatment (Fisher et al. 2007) has developed a suite of tools that can be used by distribution system managers to improve their task of disinfection. The suite encompasses design, planning, and operational tools that allow water supply companies to address factors that may enhance the performance of disinfection systems. Among the tools are process-based ones that are used to predict the decay of chlorine or chloramines in water distribution systems.

The above-described tools incorporate an improved disinfection decay model, which objectively separates two key processes: bulk decay, in which the disinfectant is consumed by reactions with NOM in the treated water, and wall decay, in which the disinfectant is consumed by reactions with biofilms, corrosion products, and sediments that line pipes. Coupled to an existing hydraulic water distribution system model (e.g., EPANET), the improved decay model allows accurate simulation of disinfectant residuals behavior, including the impact of actions such as pipe cleaning or re-chlorination. Process-based tools also allow management teams to perform scenario modeling to assess the impact of operational changes made to their system. It also allows management to predict and evaluate the performance of modified or newly designed systems. These activities can assist in the evaluation of risks (exposure to hazards) and preventive measures, both of which are part of the Framework for Management of Drinking Water Quality.

The Disinfection Management Tool (DMTool) is a process-based model that may be used for several purposes: to establish dosing and booster disinfection levels for chlorine and chloramines, to predict free and total chlorine levels, to predict levels of THMs and HAAs, and to improve the understanding of disinfection in distribution systems or their operation. This model also allows users to account for biofilms, compensate for temperature effects, locate secondary chlorinators and reduce customer complaints. In essence, The DMTool has a potential role for evaluating risks and developing preventive measures that are associated with chlorine decay and disinfection by-product formation. The task of developing data-based tools, designed to assist water supply companies to improve control of disinfection in distribution systems is clearly underway. Such disinfection control tools, also collectively known as the Disinfection Toolkit (DrCT®), consist of several tools for predicting disinfectant residual concentration, and can be used to provide advice on what the optimal doses are to accomplish target residual concentrations at key network locations.

6.5 Disinfection Profiling and Benchmarking

The practice of disinfection profiling and benchmarking is one of the requirements under the USEPA Interim Enhanced Surface Water Treatment Rule (IESWTR), promulgated on December 16, 1998. The main purpose of this practice is to ensure that microbial inactivation is not significantly reduced from implementation of the Stage 1 Disinfectant and Disinfection By-product Rule (DBPR) that was also promulgated on December 16, 1998. To meet the terms of the Stage 1 DBPR, water utilities may make significant modifications to their existing disinfection practices. Therefore, it is important for water utilities to understand the impact of these changes on microbial protection (USEPA 1999). As required under the IESWTR, a disinfection profile must be developed for a period between 1 and 3 years, depending on the availability and quality of existing data. For disinfection profiling and benchmarking, the CT value approach is used to work out the log inactivation of microbial pathogens of concern (e.g., *Giardia* or viruses) achieved during water treatment.

Once the CT value required to achieve a 3-log inactivation of *Giardia* and/or 4-log inactivation of viruses has been calculated, the actual plant CT value must be determined. From the daily estimated plant log inactivation data, a disinfection profile can then be created. Using the daily plant log inactivation records, the average log inactivation for each calendar month is subsequently calculated. The lowest monthly average log inactivation values for each 12-month period are then averaged to determine the benchmark. If 1 year of data is available, the lowest monthly average log inactivation is the disinfection benchmark.

7 Summary

Drinking water is the most important single source of human exposure to gastroenteric diseases, mainly as a result of the ingestion of microbial contaminated water. Waterborne microbial agents that pose a health risk to humans include enteropathogenic bacteria, viruses, and protozoa. Therefore, properly assessing whether these hazardous agents enter drinking water supplies, and if they do, whether they are disinfected adequately, are undoubtedly aspects critical to protecting public health. As new pathogens emerge, monitoring for relevant indicator microorganisms (e.g., process microbial indicators, fecal indicators, and index and model organisms) is crucial to ensuring drinking water safety. Another crucially important step to maintaining public health is implementing Water Safety Plans (WSPs), as is recommended by the current WHO Guidelines for Drinking Water Quality. Good WSPs include creating health-based targets that aim to reduce microbial risks and adverse health effects to which a population is exposed through drinking water.

The use of disinfectants to inactivate microbial pathogens in drinking water has played a central role in reducing the incidence of waterborne diseases and is considered to be among the most successful interventions for preserving and promoting public health. Chlorine-based disinfectants are the most commonly used disinfectants and are cheap and easy to use. Free chlorine is an effective disinfectant for bacteria and viruses; however, it is not always effective against *C. parvum* and *G. lamblia*. Another limitation of using chlorination is that it produces disinfection by-products (DBPs), which pose potential health risks of their own. Currently, most drinking water regulations aggressively address DBP problems in public water distribution systems. The DBPs of most concern include the trihalomethanes (THMs), the haloacetic acids (HAAs), bromate, and chlorite. However, in the latest edition of the WHO Guidelines for Drinking Water Quality, it is recommended that water disinfection should never be compromised by attempting to control DBPs. The reason for this is that the risks of human illness and death from pathogens in drinking water are much greater than the risks from exposure to disinfectants and disinfection by-products. Nevertheless, if DBP levels exceed regulatory limits, strategies should focus on eliminating organic impurities that foster their formation, without compromising disinfection.

As alternatives to chlorine, disinfectants such as chloramines, ozone, chlorine dioxide, and UV disinfection are gaining popularity. Chlorine and each of these disinfectants have individual advantage and disadvantage in terms of cost, efficacy, stability, ease of application, and nature of disinfectant by-products (DBPs). Based on efficiency, ozone is the most efficient disinfectant for inactivating bacteria, viruses, and protozoa. In contrast, chloramines are the least efficient and are not recommended for use as primary disinfectants. Chloramines are favored for secondary water disinfection, because they react more slowly than chlorine and are more persistent in distribution systems. In addition, chloramines produce lower DBP levels than does chlorine, although microbial activity in the distribution system may produce nitrate from monochloramine, when it is used as a residual disinfectant.

Achieving the required levels of water quality, particularly microbial inactivation levels, while minimizing DBP formation requires the application of proper risk and disinfection management protocols. In addition, the failure of conventional treatment processes to eliminate critical waterborne pathogens in drinking water demand that improved and/or new disinfection technologies be developed. Recent research has disclosed that nanotechnology may offer solutions in this area, through the use of nanosorbents, nanocatalysts, bioactive nanoparticles, nanostructured catalytic membranes, and nanoparticle-enhanced filtration.

Acknowledgments The authors would like to thank the editor and reviewers for their thorough review of our manuscript. Comments and recommendations provided were both insightful and helpful. The authors are also grateful to colleagues for their guidance and support and also Rand Water for their financial assistance.

References

Acero J, Rodriguez E, Meriluoto J (2005) Kinetics of reactions between chlorine and the cyanobacterial toxins microcystins. Water Res 39:1628–1638

Adam RD (2001) Biology of *Giardia lamblia*. Clin Microbiol Rev 14:447–475

Adham SS, Jacangelo JG (1994) Assessing the reliability of low pressure membrane systems for microbial removal. The National Conference on Environmental Engineering, Boulder, CO

Aieta EM, Berg DJ (1986) A review of chlorine dioxide in drinking water treatment. J Am Water Works Assoc 786:62–72

Akasaka T, Watari F (2009) Capture of bacteria by flexible carbon nanotubes. Acta Biomater 5:607–612

Albuquerque JEC, Mendez MOA, Coutinho ADR, Franco TT (2008) Removal of cyanobacterial toxins from drinking water by adsorption on activated carbon fibers. Mater Res 11:370–380

Antoniou P, Hamilton J, Koopman B, Jain R, Holloway B, Lyberatos G, Svoronos S (1990) Effect of temperature and pH on the effective maximum specific growth rate of nitrifying bacteria. Water Res 24(1):97–101

Arevalo JM (2007) Modelling free chlorine and chloramines decay in a pilot distribution system. PhD thesis, University of Central Florida

Ashbolt NJ (2004) Microbial contamination of drinking water and disease outcomes in developing regions. Toxicology 198:229–238

Ashbolt NJ, Grabow WOK, Snozz M (2001) Indicators of microbial water quality. In: Fewtrell L, Bartaram J (eds) World Health Organization. Water quality: guidelines, standards and health. IWA Publishing, London, pp 289–316

AWWA (1980) AWWA seminar on water disinfection with ozone, chloramines, or chlorine dioxide, Atlanta, Georgia, water quality division. American Water Works Association, Denver, CO. ISBN 0898672449

AWWA (1990) Water quality and treatment. In: Pontius FW (ed) A handbook of community water supplies. McGraw-Hill, New York

AWWA (2001a) Water quality control in distribution system. A Policy Statement Adopted by the AWWA Board of Directors

AWWA (2006) Waterborne pathogens. AWWA manual M48. In: Behner B, Christensen M (eds) 2nd edn, Glacier Publishing Services Inc. American Water Works Association, Denver, CO. ISBN: 1583214038

AWWA (2008) Committee report: disinfection survey, part 1-recent changes, current practices, and water quality. JAWWA 100(10):76–90

Backer H, Hollowell J (2000) Use of iodine for water disinfection: iodine toxicity and maximum recommended dose. Environ Health Perspect 108:679–684

Badireddy AR, Hotze EM, Chellam S, Alvarez P, Wiesner MR (2007) Inactivation of bacteriophages via photosensitization of fullerol nanoparticles. Environ Sci Technol 41(18):6627–6632

Baribeau H, Prevost M, Desjardins R, Lafrance P, Gates DJ (2002) Chlorite and chlorate ion variability in distribution systems. J Am Water Works Assoc 94(7):96–105

Bartram J, Cotruvo J, Exner M, Fricker C, Glasmacher A (2003) Heterotrophic plate counts and drinking-water safety: the significance of HPCs for water quality and human health. WHO emerging issues in water and infectious disease series. IWA Publishing, London

Batterman S, Zhang L, Wang S (2000) Quenching of chlorination disinfection by-product formation in drinking water by hydrogen peroxide. Water Res 34(5):1652–1658

Bauman CM, vd Walt CJ, Kruger M, Pietersen J, Malan J, Haarhoff J (2002) Filter integrity asessment at Midvaal Water Company. Presented at the Water Institute of Southern Africa Biennial Conference, Durban, South Africa

Beech IB, Sunner J (2004) Biocorrosion: towards understanding interactions between biofilms and metals. Curr Opin Biotechnol 15:181–186

Bellar TA, Bienlien JJ, Kroner RC (1974) The occurrence of organohalides in chlorinated drinking water. J Am Water Works Assoc 66(12):703–706

Bercz JP, Bawa R (1986) Iodination of nutrients in the presence of chlorine based disinfectants used in drinking water treatment. Toxicol Lett 34(2–3):141–147

Bhagwatula A, Temburkar AR, Gupta R (2000) Method of disinfection other than chlorination. Proceeding – all India seminars on disinfection of rural & urban water supplies, The Institution of Engineers (India). pp 85–89

Bhole AG (2000) Potassium permanganate treatment – an alternative for prechlorination. Proceeding on all India seminar on disinfection of rural & urban water supplies, The Institution of Engineers (India). pp 71–76

Blackburn BG, Craun GF, Yoder JS, Hill V, Calderon RL, Chen N, Lee SH, Levy DA, Beach MJ (2004) Surveillance for waterborne-disease outbreaks associated with drinking water–United States, 2001–2002. Morbid Mortal Weekly Rep 53(SS-8):23–45

Blaser MJ (1997) Epidemiologic and clinical features of *Campylobacter jejuni* infections. J Infect Dis 176(suppl 2):S103–S105

Block SS (2001) Disinfection, sterilization, and preservation. Lippincott Williams and Wilkins, New York

Bouchard DC, Williams MK, Surampalli RY (1992) Nitrate contamination of groundwater sources and potential health effect. J Am Water Works Assoc 84:85–90

Boyd D (2006) The water we drink: an international comparison of drinking water quality standards and guidelines. David Suzuki Foundation, Vancouver, BC

Bruce-Grey-Owen Sound Health Unit (2000) The investigative report on the Walkerton outbreak of gastroenteritis. URL http://www.publichealthgreybruce.on.ca/private/Reports/SPreport.htm. Accessed 23 Sept 2011

Bukhari Z, Hargy TM, Bolton JR, Dussert B, Clancy JL (1999) Medium-pressure UV light for oocyst inactivation. J Am Water Works Assoc 91(3):86–94

Camper A, Burr M, Ellis B, Butterfield P, Abernathy C (1999) Development and structure of drinking water biofilms and techniques for their study. J Appl Microbiol Symp Suppl 85:1S–12S

Camper AK (2004) Involvement of humic substances in re-growth. Int J Food Microbiol 92:355–364

Cannon B, Herbert N, Strohwald H (2008) Final effluent disinfection at Potsdam works by means of UV irradiation. Presented at the Water Institute of Southern Africa biennial conference, Durban, South Africa

Casson LW, Hunter GL, Bess JW Jr (2006) The challenge of the future: operating emerging disinfection technologies. Proceedings for WEFTEC®.06. Water Environment Foundation

Casteel MJ, Sobsey MD, Arrowood MJ (2000) Inactivation of *Cryptosporidium parvum* oocysts and other microbes in water and wastewater by eletrolytically generated mixed oxidants. Water Sci Technol 41(7):127–134

Caughran TV, Richardson SD, Thruston AD Jr, Chen PH, Collette TW, Floyd TL (1999) Identification of new drinking water disinfection byproducts formed in the presence of bromide. Book of abstracts. 217th Amercian Chemical Society national Meeting. Anaheim, CA, 21–25 March

Chambers CW, Protor CM, Kabler PW (1962) Bactericidal effects of low concentration of silver. J Am Water Works Assoc 54:208–216

Cho M, Chung H, Choi W, Yoon J (2005) Different inactivation behavior of MS-2 phage and *Escherichia coli* in TiO_2 photocatalytic disinfection. Appl Environ Microbiol 71(1):270–275

Cho M, Chung H, Yoon J (2003) Quantitative evaluation of the synergistic sequential inactivation of *Bacillus subtilis* spores with ozone followed by chlorine. Environ Sci Technol 37(10):2134–2138

Cho M, Lee Y, Choc W, Chung H, Yoon J (2006) Study of Fe(VI) species as a disinfectant: quantitative evaluation and modeling for inactivating *E.coli*. J Water Res 40:3580–3586

Chorus I, Bartrum J (eds) (1999) Toxic cyanobacteria in water: a guide to their public health consequences, monitoring and management. E & FN Spon (published on behalf of the World Health Organization), London

Clark RM, Lykins BW Jr, Block JC, Wymer LJ, Reasoner DJ (1994) Water quality changes in a simulated distribution system. J Water Suppl Res Technol Aqua 43(6):263–277

Connell GF, Routt JC, Macler B, Andrews RS (2000) Committee report: disinfection at large and medium-size systems. J Am Water Works Assoc 92(5):32–43

Cooney CM (2008) Drinking-water analysis turns up even more toxic compounds. Environ Sci Technol 42(22):8175

Corona-Vasquez B, Rennecker JL, Driedger AM, Mariñas BJ (2002) Sequential inactivation of *Cryptosporidium pavum* oocysts with chlorine dioxide followed by free chlorine or monochloramine. Water Res 36(1):178–188

CPHEEO (1999) Manual on water supply and treatment. Ministry of Urban Development, GOI, New Delhi

Craik SA, Finch GR, Bolton JR, Belosevie M (2001) Inactivation of *cryptosporidium parvum* oocysts using medium- and low-pressure ultraviolet radiation. Water Res 35(6):1387–1398

Craun GF, Nwachuku N, Calderon RL, Craun MF (2002) Outbreaks in drinking-water systems, 1991–1998. J Environ Health 65:16–25

CRC (2008) All about Australia's drinking water. Drinking Water Facts 3:1–4

Croll BT (1992) Membrane technology: the way forward. J Instit Water Environ Manag 6(2):121–129

CSV Water (2011) Extension and upgrading of the Brits WTP. http://www.csvwater.co.za/projects/ProjectSheet_BritsWTP.pdf. Accessed 16 Aug 2011

Dawson RM (1998) The toxicology of microcystins. Tetrahedron 36:953–962

Deere D, Stevens M, Davison A, Helm G, Dufour A (2001) Management strategies. In: Fewtrell L, Bartram J (eds) Water quality: guidelines, standards and health. IWA Publishing, London, pp 257–288

DEFRA (2011) A review of fungi in drinking water and the implications for human health. Final report. Reference: WD 0906. http://dwi.defra.gov.uk/research/completed-research/reports/DWI70-2-255.pdf. Accessed 04 Jan 2012

De Roos AJ, Ward MH, Lynch CF, Cantor KP (2003) Nitrate in public water supplies and the risk of colon and rectum cancers. Epidemiology 14(6):640–649

Dechesne M, Soyeux E, Loret JF, Westrell T, Stenstrom TA, Gornik V, Koch C, Exner M, Stranger M, Agutter P, Lake R, Roser D, Ashbolt N, Dullemont Y, Hijnen W, Medema GJ (2006) Pathogens in source water. Report of the Microrisk Project to the European Commission (contract EVK1-CT-2002-00123)

De Gusseme B, Sintubin L, Baert L, Thibo E, Hennebel T, Vermeulen G, Uyttendaele M, Verstraete W, Boon N (2010) Biogenic silver for disinfection of water contaminated with viruses. Appl Environ Microbiol 76(4):1082–1087

Deng S, Upadhyayula VKK, Smith GB, Mitchell MC (2008) Adsorption equilibrium and kinetics of microorganisms on single walled carbon nanotubes. IEEE Sens 8:954–962

Dewettinck T, Van Houtte E, Geenens D, Van Hege K, Verstraete W (2001) HACCP to guarantee safe water reuse and drinking water production – a case study. Water Sci Technol 43(12):31–38

DiGiano FA, Zhang W (2005) Uncertainty analysis in a mechanistic model of bacterial regrowth in distribution systems. Environ Sci Technol 38:5925–5931

Dlyamandoglu V, Selleck RE (1992) Reactions and products of chloramination. Environ Sci Technol 26(4):808

DOE (2006) Department of Environment, Newfoundland, Water Resources Management

Doshi PE, Grayman WM, Guastella D (2003) Field testing the chlorine wall demand in distribution mains. In: Proceedings of the 2003 annual conference of the american water works association, American Water Works Association, Denver, CO. pp 1–10

Driedger AM, Rennecker JL, Mariñas BJ (2000) Sequential inactivation of Cryptosporidium pavum oocysts with ozone and free chlorine. Water Res 34(14):3591–3597

Drown T (1894) Electrical purification of water. J N Engl Water Works Assoc V8:183

DuPont HL, Chappell CL, Sterling CR, Okhuysen PC, Rose JB, Jakubowski W (1995) The infectivity of Cryptosporidium parvum in healthy volunteers. N Engl J Med 332(13):855–859

Edwards D, Donn A, Meadowcroft C (2001) Membrane solution to a "significant risk" Cryptosporidium groundwater source. Dasalination 137(1–3):193–198

Eisnor JD, Gagnon GA (2004) Impact of secondary disinfection on corrosion in a model water distribution system. J Water Suppl Res Technol (Aqua) 53(7):441–452

Emtiazi F, Schwartz T, Marten SM, Krolla-Sidenstein P, Obst U (2004) Investigation of natural biofilms formed during the production of drinking water from surface water embankment filtration. Water Res 38:1197–1206

EU (1998) Council Directive 98/83/EC. Quality of water intended for human consumption

EU (2006) The new groundwater directive (2006/118/EC). European Commission

Euro Chlor (2012) Chlorine as a disinfectant. http://www.cloro.info/what-is-chlorine/chlorine-as-a-water-disinfectant. Accessed 06 July 2011

European Commission (2008) The quality of drinking water in the European Union. Synthesis report on the quality of drinking water in the member states of the European Union in the period 1999–2001. Directive 80/778/EEC, 14 April 2008

Ewald PW (1996) Guarding against the most dangerous emerging pathogens: insights from evolutionary biology. Emerg Infect Dis 2(4):245–257

Falconer IR, Bartram J, Chorus I, Kuiper-Goodman T, Utkilen H, Burch M, Codd GA (1999) Safe levels and safe practices. In: Chorus I, Bartram J (eds) Toxic cyanobacteria in water: a guide to their public health consequences, monitoring and management. E&FN Spon, London, pp 155–178

Falkinham JO III, Norton CD, LeChevallier MW (2001) Factors influencing numbers of *Mycobacterium avium*, *Mycobacterium intracellulare*, and other Mycobacteria in drinking water distribution systems. Appl Environ Microbiol 67:1225–1231

Fane AG, Tang CY, Wang V (2011) Membrane technology for water: microfiltration, ultrafilteration, nanofiltration, and reverse osmosis, treatise on water science, Chapter 4.11, pp 301–335

Federation of Canadian Municipalities and National Research Council (2003) Wastewater treatment plant optimization. Issue no 1. ISBN 1-897094-32-9

Fields BS, Benson RF, Besser RE (2002) Legionella and Legionnaires' disease: 25 years of investigation. Clin Microbiol Rev 15:506–526

Fisher I, Kasti G, Tam T, Ye S (2007) Consolidation of disinfection management tools for distribution systems. The Cooperative Research Centre for Water Quality Treatment, Research report 34

Fliermans CB (1996) Ecology of Legionella: from data to knowledge with a little wisdom. Microb Ecol 32:203–228

Freeman SDN, Logsdon GS, Harris AT, Moles AD (1996) Evaluation of microfiltration performance with Bacillus spore, particle count and particle index measurements. The AWWA Annual Conference, Toronto, Ontario. pp 1013–1039

Genthe B, Kfir R (1995) Studies on microbiological drinking water quality guidelines. Water Research Commission Report, 469/1/95

Gerba CP, Haas CN (1988) Assessment of risks associated with enteric viruses in contaminated drinking water. ASTM Spec Tech Publ 976:489–494

Gerba CP, Naranjo JE, Hasan MN (1997) Evaluation of a combined portable reverse osmosis and iodine resin drinking water treatment systems for control of enteric pathogens. J Environ Health Sci A 32(8):2337–2354

Glaze WH (1987) Drinking-water treatment with ozone. Environ Sci Technol 21(3):224–230

Glaze WH, Weinberg HS, Cavanagh JE (1993) Evaluating the formation of brominated DBPs during ozonation. J Am Water Works Assoc 85(1):96–106

Global Research Coalition (2009) Annual report 2009–2010. Water Quality Research Australia Limited

Gomez M, de la Rua A, Garralon G, Plaza F, Hontoria E, Gomez MA (2006) Urban wastewater disinfection by filtration technologies. Desalination 190:16–28

Goodyer L, Behrens RH (2000) Safety of iodine based water sterilization for travelers. J Travel Med 7:38

Grayman WM, Rossman LA, Gill MA, Li Y, Guastella DE (2002) Measuring and modeling disinfectant wall demand in metallic pipes". In: Proceedings of the EWRI conference on water resources planning and management, Reston VA, ASCE

Guo H, Wyart Y, Perot J, Nauleau F, Moulin P (2010) Low-pressure membrane integrity tests for drinking water treatment: a review. Water Res 44:41–57

Haas CN (1999) Disinfection. In: Letterman RD (ed) Quality and treatment: a handbook of community water supplies. McGraw-Hill, New York, pp 877–932

Haas CN, Trussell RR (1998) Frameworks for assessing reliability of multiple, independent barriers in potable water reuse. Water Sci & Technol 38(6):1–8

Hageskal G, Knutsen AK, Gaustad P, de Hoog GS, Skaar I (2006) The diversity and significance of mold species in Norwegian drinking water. Appl Environ Microbiol 72(12):7586–7593

Haider S, Naithani V, Viswanathan PN, Kakkar P (2003) Cyanobacterial toxins: a growing environmental concern. Chemosphere 52:1–21

Hazen and Sawyer (1992) Disinfection alternatives for safe drinking water. Van Nostrand Reinhold, New York

Health Canada (1996) A one-year survey of halogenated disinfection by-products in the distribution system of treatment plants using three different disinfection processes, Minister of Public Works and Government Services Canada, H46-2/96-206E

Health Canada (2006) Guidelines for Canadian drinking water quality, 6th edn. Minister of Supply and Services Canada, Ottawa, Canada (Catalogue No. H48-10/1996E)

Health Canada (2009) Guidelines for Canadian drinking water quality guideline technical document chlorine. Water, Air and Climate Change Bureau, Healthy Environments and Consumer Safety Branch, Health Canada, Ottawa, Ontario (Catalogue No. H128-1/09-588E)

Health Canada (2010) Guidelines for Canadian Drinking water quality-summary table. http://www.hc-sc.gc.ca/ewh-semt/alt_formats/hecs-sesc/pdf/pubs/water-eau/2010-sum_guide-res_recom/sum_guide-res_recom-eng.pdf. Accessed 12 Oct 2011

Heidarpour F, Wak G, Fakhru'l-Razi A, Sobri S, Torabian A, Heydarpour V, Zargar M (2011) New trends in microbiological water treatment. Digest J Nanomater Biostruct 6(2):791–802

Hirata T, Hashimoto A (1998) Experimental assessment of the efficacy of microfiltration and ultrafiltration for Cryptosporidium removal. Water Sci Technol 38(12):103–107

Hoff JC, Akin EW (1986) Microbial resistance to disinfectants: mechanisms and significance. Environ Health Perspect 69:7–13

Hoff JC (1992) The Relationship of Turbidity to Disinfection of Potable Water. In: Hendricks CW (ed) Evaluation of the microbiology standards for drinking water. United States Environmental Protection Agency, Cincinnati, OH, pp 103–117

Hirtle LE (2008) Exploring pretreatments for the solar water disinfection (SODIS) process. Master's thesis, University of Waterloo, Waterloo, Canada

Howard G, Pedley S, Barret M, Nalubega M, Johal K (2003) Risk factors contribution to microbiological contamination of shallow groundwater in Kampala, Uganda. Water Res 37:3421–3429

Howard G, Pedley S, Tibatemwa S (2006) Quantitative microbial risk assessment to estimate health risks attributable to water supply: can the technique be applied in developing countries with limited dat? J Water Health 4(1):49–65

Hrudey SE, Payment P, Huck PM, Gillham RW, Hrudey EJ (2003) A fatal waterborne disease epidemic in Walkerton, Ontario: comparison with other waterborne outbreaks in the developed world. Water Sci Technol 47(3):7–14

Hrudey SE, Hrudey EJ (2004) Safe drinking water – lessons from recent outbreaks in affluent nations. IWA Publishing, London

Huang SS, Labus BJ, Samuel MC, Wan DT, Reingold AL (2002) Antibiotic resistance patterns of bacterial isolates from blood in San Francisco County, California, 1996–1999. Emerg Infect Dis 8(2):195–201

Hunter PR (1997) Waterborne diseases epidemiology and ecology, 1st edn. Wiley, Chichester. ISBN: 0471-96646-0

Hunter PR, Syed Q (2001) Community surveys of self-reported diarrhea can dramatically overestimate in size of outbreaks of waterborne cryptosporidiosis. Water Sci & Technol 43:27–30

Hunter PR, Payment P, Ashbolt N, Bartram J (2003) Assessment of risk. WHO, Geneva

Hunterwater (2009) Chlorination of water supplies. Factsheet. http://www.hunterwater.com.au/Resources/Documents/Fact-Sheets/Water-Quality/Chlorination-of-Water-Supplies.pdf. Accessed 13 Aug 2011

Hydes O (1999) European regulations on residual disinfectant. J Am Water Works Assoc 91(1):70–74

ILSI (1996) a conceptual framework to assess the risks of human disease following exposure to pathogens. Risk Anal 16:841–848

Isaac RA, Morris JC (1983) Transfer of active chlorine from chloramine to nitrogenous organic compounds. Environ Sci Technol 17:738–742

Jacangelo JG, Adham SA, Laîné JM (1995) Mechanism of Cryptosporidium parvum, Giardia muris and MS2 virus removal by MF and UF. J Am Water Works Assoc 87(9):107–121

Jacangelo J, Adham S, Laine JM (1997), Membrane filtration for microbial removal. Report No. 90715. American Water Works Association Research Foundation, Denver, CO

Jacangelo JG, Loughran P, Petrik B, Simpson D, McIlroy C (2003) Removal of enteric viruses and selected microbial indicators by UV irradiation of secondary effluent. Water Sci Technol 47(9):193–198

Jang H (2009) Organic chloramines formation and its disinfection efficacy. PhD thesis, Arizona State University, 157 p

Jee KY, Lee C, Cho M, Yoon J (2008) Enhanced inactivation of *E. coli* and MS-2 phage by silver ions combined with UV-A and visible light irradiation. Water Res 42:356–362

Jesperson K (2004) Emerging and re-emerging pathogens: compelling reasons to protect drinking water. On Tap Fall 2004

Joyce TM, McGuigan KG, Elmore-Meegan M, Conroy RM (1996) Inactivation of faecal bacteria in drinking water by solar heating. Appl Environ Microbiol 62:399–402

Junli H, Li W, Nanqi R, Fang M, Juli S (1997) Disinfection effect of chlorine dioxide on bacteria in water. Water Res 31:607–613

Kaminski JC (1994) Cryptosporidium and the public water supply. N Engl J Med 331:1529

Kang S, Pinault M, Pfefferle LD, Elimelech M (2007) Single walled carbon nanotubes exhibit strong antimicrobial activity. Langmuir 23:8670–8673

Kawamura S (2000) Integrated design and operation of water treatment facilities, 2nd edn. Wiley, New York. pp 229–237, 543–552

Kelley J, Kinsey GC, Paterson RRM, Pitchers R (2001) Identification and control of fungi in distribution systems. AWWA Research Foundation, Denver, CO

Kelley J, Paterson R, Kinsey G, Pitchers R, Rossmoore H (1997) Identification, significance and control of fungi in water distribution systems. Water technology conference proceedings, 9–12 November 1997, Denver, CO, USA. Public American Water Works Association

Kim J, Cho M, Oh B, Choi S, Yoon J (2004) Control of bacterial growth in water using synthesized inorganic disinfectant. Chemosphere 55:775–780

Kirmeyer GJ, Foust GW, Pierson GL, Simmler JJ, LeChevallier MW (1993) Optimizing chloramine treatment. AwwaRF and AWWA, Denver, CO

Kirmeyer GJ, Odell LH, Jacangelo J, Wilczak A, Wolfe R (1995) Nitrification occurrence and control in chloraminated water systems. AwwaRF and AWWA, Denver, CO

Kusumawardaningsih Y (2010) A preliminary review of some drinking water quality's guidelines and standards in the world. Jurnal Kompetensi Teknik 1(2):75–82

LeChevallier MW (2006) Assessing distribution system intergrity: the case for maintaining a disinfectant residual. American Water. http://www.odwac.gov.on.ca/Alternative_Approaches_to_Disinfection/Presentations/4_LeChevallier.pdf. Accessed 01 July 2011

LeChevallier MW, Au K (2004) Water treatment and pathogen control: process efficiency in achieving safe drinking water. IWA Publishing, London

Lenntech BV (2011) Water disinfection application standards (for EU). http://www.lenntech.com/processes/disinfection/regulation-eu/eu-water-disinfection-regulation.htm. Accessed 03 June 2011

Liang JL, Dziuban EJ, Craun GF, Hill V, Moore MR, Gelting RJ, Calderon RL, Beach MJ, Roy SL (2006) Surveillance for waterborne disease outbreaks associated with drinking water and water not intended for drinking—United States, 2003–2004. Morbid Mortal Weekly Rep 55(12):31–65

Liau SY, Read DC, Pugh WJ, Furr JR, Russell AD (1997) Interaction of silver-nitrate with readily identifiable groups: relationship to the antibacterial action of silver ions. Lett Appl Microbiol 25:279–283

Lipponen MT, Martikainen PJ, Vasara RE, Servomaa K, Zacheus O, Kontro MH (2004) Occurrence of nitrifiers and diversity of ammonia-oxidizing bacteria in developing drinking water biofilms. Water Res 38:4424–4434

Liu Y, Mou S, Heberling S (2002) Determination of trace level bromate and perchlorate in drinking water by ion chromatography with an evaporative preconcentration technique. J Chromatogr A 956:85–91

Loret JF, Robert S, Thomas V, Cooper AJ, McCoy WF, Levi Y (2005) Comparison of disinfectants for biofilm, protozoa and Legionella control. J Water Health 34:423–433

Lund E (1963) Oxidative inactivation of poliovirus at different temperatures. Arch Virol 13(4):375–386

Lyon DY, Adams LK, Falkner JC, Alvarez PJJ (2006) Antibacterial activity of fullerene water suspensions: effects of preparation method and particle size. Environ Sci Technol 40(14):4360–4366

MacKenzie WR, Hoxie NJ, Proctor ME, Gradus MS, Blair KA, Peterson DE, Kazmierczak JJ, Addiss DG, Fox KR, Rose JB, Davis JP (1994) A massive outbreak in Milwaukee of cryptosporidium infection transmitted through the public water supply. N Engl J Med 331:161–167

MacPherson G, Lombard P (2006) Bitou municipality: ozonation at the central water purification works at Plettenberg Bay. Presented at the Water Institute of Southern Africa biennial conference, Durban, South Africa

Mains C (2008) Biofilm control in distribution systems. Tech Brief 8(2):1–4

Malley JP, Shaw JP, Ropp JD (1995) Evaluation of by-products produced by treatment of groundwaters with ultraviolet irradiation. AWWA Research Foundation Report No. 90685. Denver CO

Mani SK, Kanjur R, Isaac S, Singh B, Reed RH (2006) Comparative effectiveness of solar disinfection using small-scale batch reactors with reflective absorptive and transmissive rear surfaces. J Water Res 40:721–722

Maul A, El-Shaarawi AH, Block JC (1985) Heterotrophic bacteria in water distribution systems: I. Spatial and temporal variation. Sci Total Environ 44:201–214

McAmis JW (1927) Prevention of phenol taste with ammonia. J Am Water Works Assoc 17:341–450

McDonnell G, Russell AD (1999) Antiseptics and disinfectants: activity, action and resistance. Clin Microbiol Rev 12(1):147–179

McVay RD (2009) Chloramine production and monitoring in Florida's water supply systems. Florida Water Resources Journal, pp 16–26. http://www.fwrj.com/techarticles/0409%20 FWRJ_tech1.pdf. Accessed 05 May 2011

Medema GJ, Hoogenboezem W, van der Veer AJ, Ketelaars HAM, Hijnen WAM, Nobel PJ (2003) Quantitative risk assessment of Cryptosporidium in surface water treatment. Water Sci Technol 47:241–247

Merel S, Clément L, Thomas O (2010) Review state of the art on cyanotoxins in water and their behaviour towards chlorine. Toxicon 55:677–691

Miller GW, Rice RG Robson CM, Kuhn W, Wolf H (1978) An assessment of ozone and chlorine dioxide for treatment of municipal water supplies. EPA 600/8-78-018. USEPA, 1978

MIOX Literature (1995) Alternative comparison for important features in municipal plants and wells. MIOX Corporation, Los Alamos, NM

MOE (2006) Drinking Water Surveillance Program (DWSP) summary report for the years 2000–2004 on 179 municipal drinking water supply systems in Ontario. Ministry of Environment, Ontario

Momba MNB, Cloete TE, Venter SN, Kfir R (1998) Evaluation of the impact of disinfection processes on the formation of biofilms in potable surface water distribution systems. Water Sci Technol 38(8–9):283–289

Momba MNB, Obill CL, Thompson P (2009) Survey of disinfection efficiency of small drinking water treatment plants: challenges facing small water treatment plants in South Africa. Water SA 35(4):485–493

Morones JR, Elechiguerra JL, Camacho A, Holt K, Kouri JB, Ramirez JT, Yacaman MJ (2005) The bactericidal effect of silver nanoparticles. Nanotechnology 16(10):2346–2353

Mortula M, Imran S (2006). Disinfection practices and the challenges of byproducts in drinking water. INFRA 2006, Loews Le Condorde, Québec, PQ. pp 1–11

Mostafavia ST, Mehrniaa MR, Rashidib AM (2009) Preparation of nanofilter from carbon nanotubes for application in virus removal from water. Desalination 238:271–280

Mukiibi M, Sherwin N (2011) The good, bad and ugly of using chlorine for disinfecting drinking water. Water Cond Purif 53(2):54–60

Mura M, Bull TJ, Evans H, Sidi-Boumedine K, McMinn L, Rhodes G, Pickup R, Hermon-Taylor J (2006) Replication and long-term persistence of bovine and human strains of *Mycobacterium avium* subsp. Paratuberculosis within *Acanthamoeba polyphaga*. Appl Environ Microbiol 72:854–859

MWH (2005) Water treatment: principles and design, 2nd edn. Wiley, Hoboken, NJ

Administration NF (2001) Livsmedelsverkets föreskrifter om dricksvatten. SLVFS 2001:30

Niemi RM, Knuth S, Lundström K (1982) Actinomycetes and fungi in surface waters and in potable water. Appl Environ Microbiol 43(2):378–388

Nicholson BC, Rositano J, Burch MD (1994) Destruction of cyanobacterial peptide hepatotoxins by chlorine and chloramine. Water Res 28(6):1297–1303

NHMRC (2011) Australian drinking water guidelines. Commonwealth of Australia 2011. ISBN Online: 1864965118, Published: October 2011

NHMRC and NRMMC (2004) Australian drinking water quality guidelines. In: NHMRC, NRMMC (eds) EH19. National Health and Medical Research Council, Canberra. pp 1–615

Nichols RAB, Campbell BM, Smith HV (2003) Identification of *Cryptosporidium spp.* oocysts in United Kingdom noncarbonated natural mineral waters and drinking waters by using a modified nested PCR-restriction fragment length polymorphism assay. Appl Environ Microbiol 69:4183–4189

Norton CD, LeChevallier MW (2000) A pilot study of bacteriological population changes through potable treatment and distribution. Appl Environ Microbiol 66(1):268–276

Odell LH, Kirmeyer GJ, Wilczak A, Jacangelo JG, Marcinko JP, Wolfe RL (1996) Controlling nitrification in chloraminated systems. J AWWA 88(7):86–98

Oldenburg PS, Regan JM, Harrington GW, Noguera DR (2002) Kinetics of nitrosomonas europaea inactivation by chloramine. J AWWA 94(10):100–110

Pagnier I, Merchat M, Raoult D, La Scola B (2009) Emerging *Mycobacteria* spp. in cooling towers. Emerg Infect Dis 15(1):121–122

Park SR, Mackay WG, Reid DC (2001) *Helicobacter* sp. recovered from drinking water biofilm sampled from a water distribution system. Water Res 35:1624–1626

Patermarakis G, Fountoukidis E (1990) Disinfection of water by electrochemical treatment. Water Res 24(12):1491–1496

Paterson RRM, Hageskal G, Skaar I, Lima N (2009) Occurrence, problems, analysis and removal of filamentous fungi in drinking water. In: De Costa P, Bezerra P (eds) Fungicides: chemistry, environmental impacts and health effects. Nova Science Publishers Inc, Hauppauge, NY

Patwardhan AR (1990) Our water, our life. CAPART, New Delhi

Pavlov DN, de Wet CME, Grabow WOK, Ehlers MM (2004) Potentially pathogenic features of heterotrophic plate count bacteria isolated from treated and untreated drinking water. Int J Food Microbiol 92:275–287

PCI Africa (2005) Phase 5 - extensions to the Temba water supply scheme. http://www.pciafrica.com/example-temba.html. Accessed 16 Aug 2011

Pedahzur R, Katzenelson D, Barnea N, Lev O, Shuval H, Fattal B, Ulitzur S (2000) The efficacy of long-lasting residual drinking water disinfectants based on hydrogen peroxide and silver. Water Sci Technol 42(1–2):293–298

Pedley S, Bartram J, Rees G, Dufour A, Cotruvo JAE (2004) Pathogenic Mycobacteria in water - a guide to public health consequences, monitoring and management. Emerging issues in water and infectious disease series, World Health Organization titles with IWA Publishing. pp 1–222

Peldszus S, Andrews SA, Souza R, Smith F, Douglas I, Boulton J, Huck PM (2004) Effect of medium pressure UV irradiation on bromate concentration in drinking water: a pilot scale study. Wat Res 38:211–217

Peltier S, Cotte E, Gatel D, Herremans L, Cavard J (2003) Nanofiltration improvements of water quality in a large distribution system. Water Sci Technol Water Suppl 3:193–200

Percival S, Walker J, Hunter PR (2000) Microbiological aspects of biofilms in drinking water. CRC, Boca Raton, FL

Pfaller MA, Pappas PG, Winguard JR (2006) Invasive fungal pathogens: current epidemiological trends. Clin Infect Dis 43:S3–S14

Pianta R, Boller M, Urfer D, Chappaz A, Gmünder A (2000) Costs of conventional vs. membrane treatment for carstic spring water. Desalination 131:245–255

Pieterse JWD, Kruger MJF, Willemse GA (1993) Treatment of eutrophic water at the Western Transvaal Regional Water Company (WTRWC). Presented at the Water Institute of Southern Africa biennial conference, Durban, South Africa

Pintar KDM, Slawson RM (2003) Effect of temperature and disinfection strategies on ammonia-oxidizing bacteria in a bench-scale drinking water distribution system. Water Res 37:1805–1817

Pintar KDM, Anderson WB, Slawson RM, Smith EF, Huck PM (2005) Assessment of a distribution system nitrification critical threshold concept. J AWWA 97(7):116–129

Post GB, Atherholt TB, Cohn PD (2011) Water quality and treatment: a handbook on drinking water, 6th edn. American Water Works Association, Denver, CO

Powell JC (2000) Performance of various kinetic models for chlorine decay. J Water Resour Plann Manage 126:13–20

Pryor MJ, Naidoo PS Bahrs P, Freese SD (2002) Ozone experience at Umgeni Water. Presented at the biennial conference of the Water Institute of Southern Africa

Qi L, Xu Z, Jiang X, Hu C, Zou X (2004) Preparation and antibacterial activity of chitosan nano-particles. Carbohydr Res 339(16):2693–2700

Race J (1918) Chlorination of water. Wiley, New York

Rajagopaul R, Mbongwa NW, Nadan C (2008) Guidelines for the selection and effective use of ozone in water treatment. WRC Report No. 1596/01/08. Water Research Commission, Pretoria

Rakness KL, Stover EL, Krenek DL (1984) Design, start-up and operation of an ozone disinfection unit. J Water Pollut Control Fed 56:1152–1159

Regan JM, Harrington GW, Noguera DR (2002) Ammonia- and nitrite-oxidizing bacterial communities in a pilot-scale chloraminated drinking water distribution system. Appl Environ Microbiol 68(1):73–81

Regan JM, Harrington GW, Baribeau H, Leon RD, Noguera DR (2003) Diversity of nitrifying bacteria in full-scale chloraminated distribution systems. Water Res 37:197–205

Regli S, Rose JB, Haas CN, Gerba CP (1991) Modeling the risk from Giardia and Viruses in drinking-water. J Am Water Works Assoc 83(11):76–84

Richardson SD, Simmons JE, Rice G (2002) Disinfection byproducts: the next generation. Environ Sci Technol 36:198A–205A

Richardson S (2003) Water analysis emerging contaminants and current issues. Anal Chem 75(12):2831–2857

Richardson MD, Warnock DW (2003) Fungal infection: diagnosis and management. Blackwell Publishing, Oxford, p 307

Ridgway HF, Means EG, Olson BH (1981) Iron bacteria in drinking-water distribution systems: elemental analysis of Gallionella stalks using X-ray energy-dispersive microanalysis. Appl Environ Microbiol 41:288–292

Riffard S, Springthorpe S, Filion L, Sattar SA (2004) Occurrence of Legionella in Groundwater. American Water Works Association Research Foundation Report 90985F, Denver, CO

Rittmann BE, Snoeyink VL (1984) Achieving biologically stable drinking water. J AWWA 76(10):106–114

Rochelle PA, Marshall MM, Mead JR, Johnson AM, Korich DG, Rosen JS, De LR (2002) Comparison of in vitro cell culture and a mouse assay for measuring infectivity of *Cryptosporidium parvum*. Appl Environ Microbiol 68:3809–3817

Rodriguez E, Onstad GD, Kull TPJ (2007) Oxidative elimination of cyanotoxins: comparison of ozone, chlorine, chlorine dioxide and permanganate. Water Res 41(15):3381–3393

Rogers J, Keevil CW (1992) Immunogold and fluorescein immunolabelling of Legionella pneumophila within an aquatic biofilm visualized by using episcopic differential interference contrast microscopy. Appl Environ Microbiol 58:2326–2330

Rook JJ (1974) Formation of haloforms during chlorination of natural waters. Watet Treat Exam 23(2):234–243

Rose JB, Gerba CP, Jakubowski W (1991) Survey of potable water supplies for Cryptosporidium and Giardia. Environ Sci Technol 25(8):1393–1400

Rose JB, Lisle JT, Lechevallier M (1997) Waterborne cryptosporidiosis: incidence, outbreaks and treatment strategies. In: Fayer R (ed) Cryptosporidium and cryptosporidiosis. CRC, New York, pp 93–109

Rositano J, Nicholson BC, Pieronne P (1998) Destruction of cyanobacterial toxins by ozone. Ozone Sci Eng 20:223–238

Rositano J, Newcombe G, Nicholson B, Sztajnbok P (2001) Ozonation of NOM and algal toxins in four treated waters. Water Res 35(1):23–32

Russell AD, Hugo WB (1994) Antimicrobial activity and action of silver. Prog Med Chem 31:351–370

SABS (2011) South African national standard for drinking water. SANS 241:2011 Part 1 and Part II. Online Draft Version

Sahota P, Pandove G (2010) Biomonitoring of indicator and emerging pathogens in piped drinking water in Ludhiana. Rep Opin 2(1):14–21

Salih FM (2003) Formulation of a mathematical model to predict solar water disinfection. J Water Res 37:3921–3927

SANDEC (2002) Solar water disinfection: a guide for the application of SODIS. SANDEC Report No 06/ 02

Sarin P, Snoeyink VL, Bebee J, Kriven WM, Clement JA (2001) Physico-chemical characteristics of corrosion scales in old iron pipes. Wat Res 35:2961–2969

Sarin P, Snoeyink VL, Lytle DA, Kriven WM (2004) Iron corrosion scales: models for scale growth, iron release, and colored water formation. J Environ Eng 130(4):364–373

Sassi et al. (2005) Experiments with ultraviolet light, ultrasound and ozone technologies for onboard ballast water treatment. VTT Research Notes 2313

Savage N, Diallo SM (2005) Nanomaterials and water purification: opportunities and challenges. J Nanoparticle Res 7:331–342

Schwartz T, Kohnen W, Jansen B, Obst U (2003) Detection of antibiotic-resistant bacteria and their resistance genes in wastewater, surface water, and drinking water biofilms. FEMS Microbiol Ecol 43:325–335

Sen K, Rodgers M (2004) Distribution of six virulence factors in Aeromonas species isolated from US drinking water utilities: a PCR identification. J Appl Microbiol 97:1077–1086

September SM, Els FA, Venter SN, Brozel VS (2007) Prevalence of bacterial pathogens in biofilms of drinking water distribution systems. J Water Health 5(2):219–227

Shannon MA, Bohn PW, Elimelech M, Georgladis JG, Marias BJ, Mayes AM (2008) Science and technology for water purification in the coming decades. Nature 452(20):301–310

Shin GA, Linden K, Sobsey MD (2000) Comparative inactivation of *cryptosporidium parvum* oocysts and coliphage MS2 by monochromatric UV radiation. Disinfection 6:97–102

Shin GA, Lee JK, Freeman R, Cangelosi GA (2008) Inactivation of *Mycobacterium avium* complex by UV irradiation. Appl Environ Microbiol 74:7067–7069

Shuval H, Fattal B, Nassar A, Lev O, Pedahzur R (1995) The study of the synergism between oligodynamic silver and hydrogen peroxide as a long-acting water disinfectant. Water Supply 13(2):241–251

Siddiqui MS, Amy GL, Rice RG (1995) Bromate ion formation: a critical review. J Am Water Works Assoc 87(10):58–70

Siddiqui MS, Zhai W, Amy GL, Mysore C (1996) Bromate ion removal by activated carbon. Water Res 30(7):1651–1660

Silvestry-Rodriguez N, Bright KR, Slack DC, Uhlmann DR, Gerba CP (2007) Inactivation of *Pseudomonas aeruginosa* and *Aeromonas hydrophila* by silver in tap water. J Environ Sci Health A 42:1–6

Siveka EH (1966) Reverse osmosis pilot plants. In: Merten U (ed) Desalination by reverse osmosis. MIT Press, Cambridge, pp 239–271

Skadsen J (1993) Nitrification in a distribution system. J AWWA 85(7):95–103

Skraber SJS, Gantzer C, de Roda Husman AM (2005) Pathogenic viruses in drinking-water biofilms: a public health risk? Biofilms 2:105–117

Smeets P, Rietveld L, Hijnen W, Medema G, Stenström TA (2006) Efficacy of water treatment processes. Report of the Microrisk Project to the European Commission (contract EVK1-CT-2002-00123)

Smeets PWMH, Medema GJ, van Dijk JC (2009) The Dutch secret: how to provide safe drinking water without chlorine in the Netherlands. Drink Water Eng Sci 2:1–14

Somani SB, Ingole NW (2011) Alternative approach to chlorination for disinfection of drinking water - an overview. Int J Adv Eng Res Stud 1(1):47–50

Sommer R, Cabaj A, Hirschmann G, Haider T (2008) Disinfection of drinking water by UV irradiation: basic principles - specific requirements – International implementations. Ozone Sci Eng 30(1):43–48

Spoof L, Vesterkvist P, Lindholm T, Meriluoto J (2003) Screening for cyanobacterial hepatotoxins, microcystins and nodularin in environmental water samples by reversed-phase liquid chromatography–electrospray ionisation mass spectrometry. J Chromatogr A 1020:105–119

Spoof L (2004) High-performance liquid chromatography of microcystins and nodularins, cyanobacterial peptide toxins. PhD thesis, Åbo Akademi University, Finland

Srivastava A, Srivastava ON, Talapatra S, Vajtai R, Ajayan PM (2004) Carbon nanotube filters. Nat Mater 3(9):610–614

Stout JE, Yu VL (2003) Experiences of the first 16 hospitals using copper–silver ionization for Legionella control: implications for the evaluation of other disinfection modalities. Infect Contr Hosp Epidem 24(8):563–568

Szewzyk U, Szewzyk R, Manz W, Schleifer KH (2000) Microbiological safety of drinking water. Annu Rev Microbiol 54:81–127

Tachikawa M, Tezuka M, Morita M, Isogai K, Okada S (2005) Evaluation of some halogen biocides using a microbial biofilm system. Water Res 39:4126–4132

Taylor JS, Hong SK (2000) Potable water quality and membrane technology. J Lab Med 31(10):563–568

Teunis PFM, Havelaar AH (1999) Cryptosporidium in drinking water. Evaluation of the ILSI/RSI quantitative risk assessment framework. Report No. 284 550 006 RIVM, Bilthoven, The Netherlands

Teunis PFM, Havelaar AH, Medema GJ (1994) A literature survey on the assessment of microbial risk for drinking water. RIVM Report 734301006, RIVM Bilthoven, The Netherlands

Teunis PFM, Medema GJ, Kruidenier L, Havelaar AH (1997) Assessment of the risk of infection by Cryptosporidium and Giardia in drinking water from a surface water source. Water Res 31(6):1333–1346

Teunis PFM, Rutjesa SA, Westrell T, de Roda Husman AM (2009) Characterization of drinking water treatment for virus risk assessment. Water Res 43:395–404

Teunis PFM, Kasuga F, Fazil A, Ogden ID, Rotariu O, Strachan NJC (2010) Dose–response modeling of *Salmonella* using outbreak data. Int J Food Microbiol 144(2):243–249

Thomas V, Bouchez T, Nicolas V, Robert S, Loret JF, Levi Y (2004) Amoebae in domestic water systems: resistance to disinfection treatments and implication in Legionella persistence. J Appl Microbiol 97:950–963

Thomson J, Roddick FA, Drikas M (2002) UV photooxidation facilitating biological treatment for the removal of NOM from drinking water. J Water Suppl Res Technol-Aqua 51(6):297–306

Tokajian STH, Fuad A, Hancock IC, Zalloua PA (2005) Phylogenetic assessment of heterotrophic bacteria from a water distribution system using 16S rDNA sequencing. Can J Microbiol 51:325–335

Toledo RT, Escher FE, Ayres JC (1973) Sporicidal properties of hydrogen peroxide against food spoilage organisms. Appl Microbiol 26:592–597

UNICEF (1995) The state of the world's children. Oxford University Press, Oxford

Upadhyayula VKK, Deng S, Mitchell MC, Smith GB, Nair VS, Ghoshroy S (2008a) Adsorption kinetics of *Escherichia coli* and *Staphylococcus aureus* on single walled carbon nanotube aggregates. Water Sci Technol 58:179–184

Upadhyayula VKK, Ghoshroy S, Nair VS, Smith GB, Mitchell MC, Deng S (2008b) Single walled carbon nanotubes as fluorescence biosensors for pathogen recognition in water systems. Res Lett Nanotechnol 2008:1–5

US Bureau of Reclamation and Sandia National Laboratories (2003) Desalination and water purification technology roadmap—a report of the executive committee. Water Purification Research and Development Program Report No 95, US Department of Interior, Bureau of Reclamation, January 2003

USEPA (1979) National Interim Primary Drinking Water Regulations. Control of trihalomethanes in drinking water. Final rule. Federal Register, pp 68624–68707

USEPA (1989) Drinking water; national primary drinking water regulations; filtration, disinfection; turbitity, giardia lamblia, viruses, legionella, and heterotrophic bacteria; final rule (54 FR 27486, June 29, 1989) (EPA, 1989b) 40 CFR Parts 141 and 142

USEPA (1996) The safe drinking water act amendments of 1996-strengthening protection for America's drinking water. Federal Register, 56 FR 26460

USEPA (1999) Guidance manual alternative disinfectants and oxidants. EPA 815-R-99-014. United States Environmental Protection Agency, Washington, DC

USEPA (2001b) Controlling disinfection by-products and microbial contaminants in drinking water. United States Environmental Protection Agency, EPA; 600-R-01-110

USEPA (2002) LT1ESWTR Long term first enhanced surface water treatment rule manual. Report EPA 815-F-02-001, USEPA, Washington, DC

USEPA (2006a) 40 CFR parts 9, 141 & 142 national primary drinking water regulations: long term 2 enhanced surface water treatment rule; final rule. Fed Regist 71:653–702

USEPA (2006b) 40 CFR parts 9, 141, & 142 national primary drinking water regulations: stage 2 disinfectants and disinfection byproducts rule; final rule. Fed Regist 71:388–493

USEPA (2007) In: Science Policy Council (ed) US Environmental Protection Agency nanotechnology white paper. EPA 100/B-07/001 Washington, DC

USEPA (2010) Assessment of nitrification in distribution systems in waters of elevated ammonia levels. Research project. EPA/600/F-10/001

Vaerewijck MJM, Huys G, Palomino JC, Swings J, Portaels F (2005) Mycobacteria in drinking water distribution systems: ecology and significance for human health. FEMS Microbiol Rev 29:911–934

Van der Bruggen B, Vandecasteele C, Van Gestel T, Doyen W, Leysen R (2003) A review of pressure-driven membrane processes in wastewater treatment and drinking water production. Environ Prog 22(1):46–56

Van der Walt CJ, Basson ND, Traut DF, Haarhoff J (2002) EPANET as a tool for evaluating chloramination at Sedibeng Water. Presented at the Water Institute of Southern Africa biennial conference, Durban, South Africa

van der Wielen PW, Voost S, van der Kooij D (2009) Ammonia-oxidizing bacteria and Archaea in groundwater treatment and drinking water distribution systems. Appl Environ Microbiol 75(14):4687–4695

Vander Bruggen B, Vandecasteele C (2003) Removal of pollutants from surface water and groundwater by nanofiltration overview of possible applications in the drinking water industry. Environ Pollut 122:435–445

Vikesland PJ, Ozekin K, Valentine RL (2001) Monochloramine decay in model and distribution system waters. Water Res 35(7):1766–1776

Wegelin M, Canonica S, Mechsner K, Tleischmann T, Pasaro F, Metzler A (1994) Solar water disinfection: scope of the process and analysis of radiation experiments. J Water Supply Res Technol-Aqua 43:154–169

Wei C, Lin WY, Zainal Z, Williams NE, Zhu K, Kruzic AP, Smith RL, Rajeshwar K (1994) Bactericidal activity of TiO_2 photocatalyst in aqueous media: toward a solar-assisted water disinfection system. Environ Sci Technol 28(5):934–938

Westerhoff P (2006) Chemistry and treatment of disinfection by-products in drinking water. Southwest Hydrology, pp 20–33. http://www.swhydro.arizona.edu/archive/V5_N6/feature3. pdf. Accessed 05 May 2011

Westrell T, Schönning C, Stenström TA, Ashbolt NJ (2004) QMRA (quantitative microbial risk assessment) and HACC P (hazard analysis and critical control points) for management of pathogens in wastewater and sewage sludge treatment and reuse. Water Sci Technol 50(2):23–30

Westrick JA, Szlag DC, Southwell BJ, Sinclair J (2010) A review of cyanobacteria and cyanotoxins removal/inactivation in drinking water treatment. Anal Bioanal Chem 397:1705–1714

White GC (1992) Handbook of chlorination and alternative disinfectants. Van Nostrand Reinhold, New York, NY

White GC (1999) Handbook of chlorination and alternative disinfectants. Wiley, New York

WHO (1996) WHO guidelines for drinking water quality, 2nd edn, vol 2, Health criteria and other supporting information. World Health Organization, Geneva

WHO (2000) Health for All Statistical Database. Online. European Public Health Information Network for Eastern Europe. http://www.euphin.dk/hfa/Phfa.asp

WHO (2002a) Emerging and epidemic-prone diseases. Global Defence Against the Infectious Disease Threat. World Health Organization, Geneva

WHO (2002) Managing water in the home: accelerated health gains from improved water supply. WHO/SDE/WSH/02.07

WHO (2003a) Quantifying selected major risks to health. The World Health Report 2002. World Health Organization, Geneva

WHO (2003b) Emerging issues in water and infectious disease 1–22. World Health Organization, Geneva

WHO (2005) Water safety plans: managing water from catchment to consumer. WHO, Geneva

WHO (2004) The global burden of disease 2004 update. ISBN: 978 924 156371 0

WHO (2007) Legionella and the prevention of legionellosis. World Health Organisation, Geneva, Switzerland. ISBN: 9241562978

WHO (2009) Risk assessment of *Campylobacter* spp. in broiler chickens. Available at http://www.who.int/foodsafety/publications/micro/mra11_12/en/index.html. Accessed 03 Jan 2012

WHO (2011) World health organisation guidelines for drinking water quality, 4th edn. WHO, Geneva. ISBN 978 924 154815 1

Wilczak A, Jacangelo JG, Marcinko JP, Odell LH, Kirmeyer GJ (1996) Occurrence of nitrification in chloraminated distribution systems. J AWWA 88(7):74–85

Williams DB (1953) Dechlorination linked to corrosion in water distribution systems. Water and Sewage Works 100:106–111

Williams MM, Domingo JW, Meckes MC, Kelty CA, Rochon HS (2004) Phylogenetic diversity of drinking water bacteria in a distribution system simulator. J Appl Microbiol 96:954–964

Wong KYK (2002) Ultrasound as a sole or synergistic disinfectant in drinking water. MSC thesis, Worcester Polytechnic Institute

Woolschlager J, Rittmann B, Piriou P, Kiene L, Schwartz B (2001) Using a comprehensive model to identify the major mechanisms of chloramine decay in distribution systems. Water Sci Technol Water Supply 1:103–110

Yamanaka M, Hara K, Kudo J (2005) Bactericidal actions of a silver ion solution on *Escherichia coli*, studied by energy-filtering transmission electron microscopy and proteomic analysis. Appl Environ Microbiol 71(11):7589–7593

Yan H, Gong A, He H, Zhou J, Wei Y, Lv L (2006) Adsorption of microcystins by carbon nanotubes. Chemosphere 62:142–148

Yan H, Pan G, Hua Z, Li X, Chen H (2004) Effective removal of microcystins using carbon nanotubes embedded with bacteria. Chin Sci Bull 49:1694–1698

Yang J, Harrington GW, Noguera DR (2008) Nitrification modeling in pilot-scale chloraminated drinking water distribution systems. J Environ Eng 134(9):731–742

Ye B, Wang W, Yang L, Wei J, Xueli E (2009) Factors influencing disinfection by-products formation in drinking water of six cities in China. J Hazard Mater 171:147–152

Yoder J, Roberts V, Craun GF, Hill V, Hicks L, Alexander NT, Radke V, Calderon RL, Hlavsa MC, Beach MJ, Roy SL (2008) Surveillance for waterborne disease and outbreaks associated with drinking water and water not intended for drinking–United States, 2005–2006. Morbid Mortal Weekly Rep 57(SS-9):39–69

Yoo RS, Brown DR, Pardini RJ, Bentson GD (1995) Microfiltration: a case study. J Am Water Works Assoc 87(3):38–49

Yoon JW, Hovde CJ (2008) All blood, no stool: enterohemorrhagic *Escherichia coli* O157:H7 infection. J Vet Sci 9(3):219–231

Yoshpe-Purer Y, Eylan E (1968) Disinfection of water by hydrogen peroxide. Health Lab Sci 5(4):233–238

Zhang GR, Kiene L, Wable O, Chan US, Duguet JP (1992) Modeling of chlorine residual in the water distribution network of Macao. Environ Technol 13:937–946

Zhang J, Li X, Sun X, Li Y (2005) Surface enhanced Raman scattering effects of silver colloids with different shapes. J Phys Chem B 109:12544–12548

Zhang Y, Griffin A, Edwards M (2008) Nitrification in premise plumbing: role of phosphate, pH and pipe corrosion. Environ Sci Technol 42(12):4280–4284

Zhang Y, Love N, Edwards M (2009) Nitrification in drinking water systems. Crit Rev Environ Sci Technol 39:153–220

Zhou SW, Xu FD, Li SM, Song RX, Qi S, Zhang Y, Bao YP (1997) Major origin of mutagenicity of chlorinated drinking water in China: humic acid or pollutants. Sci Total Environ 196:191–196

Index